THE EXPANDING UNIVERSE

by

SIR ARTHUR EDDINGTON

The right of the
University of Cambridge
to print and sell
all manner of books
was granted by
Henry VIII in 1534.
The University has printed
and published continuously
since 1584.

CAMBRIDGE UNIVERSITY PRESS

Cambridge

New York   New Rochelle   Melbourne   Sydney

Published by the Press Syndicate of the University of Cambridge
The Pitt Building, Trumpington Street, Cambridge CB2 1RP
32 East 57th Street, New York, NY 10022, USA
10 Stamford Road, Oakleigh, Melbourne 3166, Australia

© Cambridge University Press 1933

First published 1933
Reprinted 1933, 1944, 1946, 1952
Reissued in the Cambridge Science Classics series 1987

Printed in the United States of America

Library of Congress Cataloging-in-Publication Data

Eddington, Arthur Stanley, Sir, 1882–1944.

The expanding universe.

(Cambridge science classics)
Reprint. Originally published:
Cambridge University Press, 1933.
Includes index.
1. Expanding universe. I. Title II. Series.
QB991.E94E33 1987 523.1'8 87-23918
ISBN 0-521-34976-1 paperback

British Library Cataloging in Publication Data applied for.

PREFACE

THIS book is an expanded version of a public lecture delivered at the meeting of the International Astronomical Union at Cambridge (Massachusetts) in September 1932. It also furnished the subject-matter of a series of three addresses which were broadcast in the United States shortly afterwards.

I deal with the view now tentatively held that the whole material universe of stars and galaxies of stars is dispersing, the galaxies scattering apart so as to occupy an ever-increasing volume. But I deal with it not as an end in itself. To take an analogy from detective fiction, it is the clue not the criminal. The "hidden hand" in my story is the *cosmical constant*. In Chapter IV we see that the investigation of the expanding universe falls into line with other methods of inquiry, so that we appear to be closing down on the capture of this most elusive constant of nature.

The subject is of especial interest, since it lies at the meeting point of astronomy, relativity and wave-mechanics. Any genuine progress will have important reactions on all three.

I am treating of very recent developments; and investigations both on the theoretical and on the observational side are still in progress which are likely to teach us much more and may modify our views. It might be argued that at this stage a book is premature; but I have ventured to assume that in the mystery stories of science the reader may be as much interested in the finding and weaving together of clues for the detection of the criminal as in his final capture and execution.

Suppose then that half-way through the chase one of the blundering detectives is here summing up what has been found out and where the strongest suspicion lies. You read his discussion, not because you have confidence that he has reached the point of identifying the criminal, but because it is presumably a necessary stage in the solution of the mystery. In real life (unlike the stories) it is possible that the suspicion already rests on the right person; be that as it may, it is worth while to set forth and analyse the present state of the inquiry.

In the astronomical part of the book I follow generally the theory of Lemaître; there is a difference in our views of evolution (see p. 59), but from my point of view this is a very minor divergence. Several counter theories of the apparent recession of the nebulae have been proposed; an explanation of my general attitude towards them is given on p. 61.

The book will be found to be of uneven difficulty; and the reader who finds himself out of his depth in Chapter II may discover that the going becomes easier further on. I have endeavoured to make the explanations as simple as possible; but the book is not intended solely as a semipopular exposition, and I have not hesitated to plunge into matters of extreme difficulty when it seemed necessary for an adequate discussion of the problem.

In remembrance of the occasion of its delivery, I add here the opening words of the lecture:

This is an International Conference and I have chosen an international subject. I shall speak of the theoretical work of Einstein of Germany, de Sitter of Holland, Lemaître of Belgium. For observational data I turn to the Americans, Slipher, Hubble, Humason, recalling however that the vitally important datum of distance is found by a method which we owe to Hertzsprung of Denmark. As I must not trouble you with mathematical analysis, I have to pass over Levi-Civita of Italy whose methods and ideas we employ. But I must refer especially to the new interest which arises in the subject through its linkage to wave-mechanics; as a representative name in wave-mechanics I mention that of its originator, de Broglie of France.

My subject disperses the galaxies, but it unites the earth. May no "cosmical repulsion" intervene to sunder us!

A. S. E.

CAMBRIDGE, ENGLAND
October 1932

vii

CONTENTS

FOREWORD

by Sir William McCrea

SIR ARTHUR (STANLEY) EDDINGTON OM, FRS (1882–1944), pioneer of stellar dynamics, tester and expounder of general relativity theory, father of modern theoretical astrophysics, explorer of the foundations of physics, was one of the greatest and most influential scientists of this century.

He had long been fascinated by questions about the constants of physics, the significancce of large dimensionless numbers given by certain combinations of these, and their possible association with large-scale properties of the astronomical universe (e.g. *The Mathematical Theory of Relativity* (1923), page 167). Then about 1930 he rather stumbled upon a possible relation between these quantities and the rate of expansion of the Universe. When he wrote this in what seemed to be its simplest form, it yielded very closely the empirical rate that E.P. Hubble had just claimed to have discovered. At the same time, Eddington was seeking some 'meeting' of relativity and quantum theory. He claimed to see a connection with the foregoing relation such that, with minimal elaboration, it served to yield a theoretical prediction of the masses of the proton and electron. He was engaged in a highly sophisticated development of all this, which in due course resulted in the monumental monographs *Relativity Theory of Protons and Electrons* (1936) and *Fundamental Theory* (1946). Clever as these are in exposing fundamental problems for the whole of physics, no one has ever claimed to be convinced of the validity of the solutions they offer. No one, that is, except Eddington. He was so convinced, as early as 1932 when he wrote this small book, that he evidently considered himself justified in giving the impression that he had fully established the results, and that here he is merely suggesting why they should strike his reader as wholly plausible.

Frankly, it is a maddening production. The reader can never be quite sure when he is being invited to follow a serious argument, or when he is being – oh so delicately – conned! Either way, however, he will be entertained. The book is short, and for reasons I try to state at the end of this foreword, I am convinced that it is well worth reading.

The foreword is an attempt to supply a critical summary to which the reader may perhaps find it helpful to refer as he goes through the book. After much travail, I try to indicate what Eddington seems to be up to at the various stages of his presentation.

Eddington naturally begins with a sketch of the astronomical evidence for the recession of the galaxies. He proposes to accept this as a 'real' phenomenon in the everyday sense. Thus, any galaxy, or cluster of galaxies, is inferred to be receding from us with speed equal to its distance multiplied by a quantity H that to a first approximation does not depend upon the direction or distance to the object. Common practice here is to take distance as a number of megaparsecs, where 1 Mpc is about $3\frac{1}{4}$ million light years. The 'Hubble parameter' H is then a speed. Eddington put it at probably between 500 and 1000 km/s. Hubble's own estimate being about 550 km/s. Hubble's estimates of distance of remote objects were about ten times too small. Therefore H is now put at between 50 and 100 km/s. When this revision is taken into account we shall see that Eddington's cosmological model actually appears in a more favourable light. The recession is, of course, the manifestation of the phenomenon that gives the book its title.

Eddington then recalls Einstein's law of gravitation in its original form of the year 1915, and the fact that when applied to local phenomena anywhere in the cosmos its results are in general almost indistinguishable from those of Newton's law. On either law therefore, we might expect the system of galaxies as a whole to tend to fall together under their mutual gravitational attraction, rather than to disperse as observed. However, in 1917 Einstein had somewhat tentatively modified his law by including a *cosmical term* which had a disposable constant λ as a factor. With Eddington we shall be concerned here with the possible existence of this *cosmical constant* only as a *positive* quantity. As was said at the time, this allowed Einstein to contemplate 'finite but unbounded' space so as to avoid possible difficulties with spatial infinity. It also introduced into the theory a force of *cosmical repulsion* (p. 23) which may be negligible over small distances but important over great distances.

Eddington, who may have been the first to see this simple way of interpreting the cosmical term, seems then to say in effect, "Look at the sky; remote objects are rushing away from us; Einstein suggests there is a cosmic repulsion that increases with distance; so we are almost bound to regard each phenomenon as good evidence for the reality of the other." But this can be deceptive because there is a hidden assumption about the mean density of the Universe. Anyhow, the reader should observe that, according to the Preface, the pursuit of the value of λ is the avowed theme of the book.

Chapter II starts with a disquisition on space-curvature and on space-closure which is the one part of the presentation that seems to be uncharacteristically laboured. From a remark later in the book (p. 78) it seems that Eddington is contemplating three-dimensional space only. His treatment might have been clarified by the introduction here of the concept of a 'metrical field' that he actually uses much later (p. 106). But when all is said and done, for his own application of all this he needs only the simplest possible case of a closed 'spherical' 3-space (expanding).

Beginning then with the cosmological models of Einstein and of de Sitter (1917), Eddington supplies a brief history of relativistic cosmology and its general treatment of expanding model universes, mainly in the work of A. Friedman and of G. Lemaître.

It seems that Eddington was the first to remark that Einstein's model is simply that in which gravitational attraction is everywhere exactly counterbalanced by cosmical repulsion – in consequence of which all its properties are determined uniquely in terms of λ – and to proceed (1930) to demonstrate its instability. This is in the sense that if the model undergoes a slight expansion then that gives a slight advantage to the repulsion; that causes further expansion, and so on. Therefore such a disturbance turns it into a particular expanding model, which can of course be treated by the Friedman–Lemaître theory; it tends asymptotically to the de Sitter case. This particular model is in fact the 'Eddington model' of the present book.

Eddington says (p. 60) his discovery of the instability of the Einstein universe was what started his interest in expanding models. But it presents a nice problem: given that a universe could ever get into the Einstein state, how could it ever get out of it? For a universe cannot be disturbed 'from outside'. Eddington reviews what has been done on this problem. So far as it goes, it seems to support his own original idea that the start of the formation of condensations – embryo galaxies, say – in the initially uniform material must be what triggers the expansion. But this seems to imply that an ideally uniform state could not in fact be realized, since there would be no reason for a condensation to start anywhere in it.

Such considerations prompt what is about the only metaphysical excursion in the book. It is about the 'beginning of the Universe' (p. 55). Eddington chose a model beginning from an effectively static state largely, it seems, on aesthetic grounds. This was about as far as is possible from today's generally preferred start in a 'hot big bang' – a description equivalent to Lemaître's 'fireworks theory' (p. 59) which Eddington said left him cold (p. 56).

For his part, as soon as he appreciated the significance of expansion, Einstein himself discarded the 'cosmological term.' He said its introduction had been the worst mistake he had ever made. Eddington was to suffer some disdain for wanting to retain λ. But the various versions of 'inflationary' cosmology current in recent years have in effect re-invented λ.

Another perennial problem in the subject is that of the implications of using a model of a smoothed-out universe instead of one containing condensations of energy and stress on all scales from, say, neutrinos to superclusters of galaxies. Eddington refers to it briefly (for example p.p. 52, 53). It came up for discussion at a conference as late as 1986; it is still not decisively resolved.

Chapter III is a discussion of the Eddington model. Some properties are:

Hubble parameter $H = c/\sqrt{3}R$ (limiting value)		$= 523\,K^{-2}$ km/s/Mpc
Initial radius of 3-space	$R = 1/\sqrt{\lambda}$	$= 1.00 \times 10^{27}\,K^2$ cm
		$= 325\,K^2$ Mpc
Total mass	$M = \pi Rc^2/2G$	$= 2.13 \times 10^{55}\,K^2$ g
'Proton number'	$N = M/m_p$	$= 1.27 \times 10^{79}\,K^2$
Initial mean density	$\rho = M/2\pi^2 R^3$	$= 1.06 \times 10^{-27}\,K^{-4}$ g cm^{-3}
Cosmical constant $\lambda = 1/R^2$		$= 9.91 \times 10^{-55}\,K^{-4}$ cm^{-2}

c = light-speed, G = gravitation constant, m_p = proton mass m_e = electron mass, e = electron charge, $r_e = e^2/m_e c^2$ = classical electron radius, $R/r_e = K/N$ defines (dimensionless) K

Apart from laboratory constants, all these properties depend, as already stated, only upon λ. The formulae for R, M are Einstein's; ρ follows from these by using the volume of 3-space of radius R. The definition of N is as shown; Eddington would regard all matter as baryonic, and so N is approximately the number of protons or electrons in his universe.

According to the relativistic treatment of the expansion already mentioned, H as listed is the value of the Hubble parameter in the asymptotic state. But the theory shows that the expansion need not have proceeded far before the parameter has roughly this value; as Eddington states (p. 72), when the radius has expanded by a factor 5, the parameter is about 0.95 of its limiting value.

Without any further theoretical commitment, we can insert numerical values for all the physical constants, and obtain the listed values of the properties in terms of the dimensionless parameter K defined as shown.

If we set $K = 1$ we have precisely the results listed by Eddington (pp. 67–8), apart from slight differences evidently due to our use of slightly different values of the constants. Until it had occurred to me to proceed in this way, I had no certain idea as to how Eddington had got his values; in so far as I can follow his own account of how he did, I am still not convinced that he got them the way he says he did!

In seeking to match predictions of the model with astronomical observations made at our epoch we should need to know by how much the model should be taken to have expanded since its initial state. Eddington mentions methods that, at least in principle, may become available in order to make some inference about this. For the time being, as he writes (p.70), he will "assume for illustration that a 5-fold expansion . . . has occurred; but this number is merely a guess". It is interesting that this happens to be about the factor that one school of present day big-bang cosmologists infers for the expansion since the epoch of galaxy formation in their model.

The discussion here, and in the next chapter, becomes needlessly confused by the employment of two radii, R and another, differing by a factor $\sqrt{3}$. I venture to suggest that the reader should discount this. Whenever any symbol R occurs (with or without a suffix) he should treat it as the one listed above. He should note simply that the formulae listed then yield immediately the result for N/R quoted by Eddington on p. 112 *without the factor $\sqrt{3}$*. His reason for introducing this factor is obscure, and its inclusion can have no effect upon his general conclusions.

He then has some rather academic comments on the possibility, or impossibility, of seeing right round the universe. Of more physical import are his comments about cosmic rays in the cosmos (pp. 80–1), about the source of stellar energy (pp. 83–4), and especially about time scales (pp. 85–6).

In Eddington's time most cosmological models except his own implied an 'age of the universe' less than 2×10^9 years. This was less than half the age of the oldest rocks on Earth! (p. 86). But his own model by starting to expand slowly at first from the static case could have had a very long past life since it left that state. This was one reason why he mischievously claimed (p. 87) to be the only one with any right to believe in the expansion. His other reason was that he believed that from laboratory physics alone he could predict that the universe had to be expanding at the observed rate.

Chapter IV is an outline of Eddington's grounds for this belief. 'Outline' because he invites his readers to entertain novel concepts largely for reasons of plausibility and procedural 'good form'. He writes, "this kind of investigation can be guided by physical insight or by analytical form" (p. 110).

Up to this point, Eddington "has been contemplating . . . phenomena on the grandest scale yet imagined". In Chapter IV he wants "to turn to the other end of the scale . . . the interior of an atom" (p. 96). He asserts that the connecting link is the cosmical constant λ.

Without following Eddington for the moment, we can look at our listed properties of his model. If it has expanded significantly away from the initial state, as we have seen, it has got to exhibit a Hubble parameter close to $532\,K^{-2}$ in the usual units. The actual universe exhibits such a parameter close to 50 (as now widely accepted). If the model is applicable we conclude $K^2 \approx 10$, $K \approx 3$. If the model has expanded to 5^3 times its original volume – in accordance with Eddington's illustrative 'guess' – using this value of K, the mean density is $5^{-3}\,\rho \approx 8.5 \times 10^{-32}$ g cm^{-3}. This is a reasonable estimate of the baryonic density. With this K, the unique Eddington model may be claimed to appear quite acceptable.

The introduction of K implied no additional assumption. Whatever were taken as the observational value of H, we could find K to yield agreement. But *a priori* we should have no reason to expect a value of order of magnitude unity. Also *a priori* we should have no reason to expect the value of K that gives the 'right' Hubble parameter to be the value that gives the 'right' mean density. Still less have we any means *a priori* to assign a precise value to K. If we could do this, the Eddington model would constitute a prediction of all the large-scale properties of the universe from laboratory quantities alone.

But – a crucial 'but' – that which improves the case for the Eddington model destroys the case against its rivals! These he had very naturally disparaged when they required the age of the Universe to be no more than 2×10^9 years. If, however, the Hubble parameter is about 10 times smaller than Eddington thought, then the age given by these models is about 10 times greater. This is about the value that should be expected on astrophysical grounds.

This amounts to a glimpse of the Eddington model after Eddington's own time. Returning briefly to that time, we remind ourselves that he did not explicitly employ a parameter K. However, his page 111 amounts to the inference from a Large Numbers hypothesis that K is of order unity, followed by the assertion that $K = 1$ "looks to me to be correct". This as we have discovered, is how Eddington obtained the numerical properties listed on pp. 67–8. When he had done this he appeared to have predicted from laboratory quantities alone the rate of expansion of the

Universe H (calculated) $= 528$, H (observed) $= 550$ km/s/Mpc (pp. 67, 11), and to have the only model that overcomes the age difficulty resulting from such a rate (pp. 86–7).

Undoubtedly Eddington was convinced that he had more sophisticated reasons for his assertions than he was able to expound within the compass of this small book. Had he stopped at the suggestion that the large-scale behaviour of the cosmos might be predictable from the values of the known constants of physics, this in itself would have been a challenging idea for his time. But when he claimed to show by pure physics that H must be more than 500 in the usual units, and when Hubble appeared to keep confirming this observationally, the effect upon the development of cosmology was far-reaching. Main progress was retarded. But side effects were of great interest; had Eddington and Hubble had the right value in the 1930s, almost certainly steady-state cosmology would not have been taken seriously during about two decades after World War II, and the whole history of modern cosmology would have been incomparably less interesting.

Eddington was seeking a 'meeting' of relativity and quantum wave mechanics. He seemed to regard the relation between N, R, and r_e as a first step. But this, and the various related results we have considered, make no mention of the quantum constant h. So it is hard to see how the wave equation of the hydrogen atom, to which he makes frequent reference, can be essentially relevant.

By re-writing the electron mass m_e in a new unit depending upon h, Eddington then writes his formula for $\sqrt{N}/R$ in the form B on p. 115. He next postulates the existence of a quadratic equation for the electron and proton masses, the last two terms being as in C on p. 116 – on the indication of B – so that the required quadratic becomes F (p. 117). So far as this presentation goes, he puts the coefficient of the quadratic term equal to 10 for no stated reason save that he has this number going abegging, as the number of coefficients, in a metric for space-time. By some miracle the roots of the quadratic equation are in the ratio that agrees with the laboratory value of m_p/m_e to better than 1 per cent (but not as closely as Eddington thought (p. 119)).

Eddington believed that there must be a modified Schrödinger wave equation for the hydrogen atom, obtained by replacing the mass in his quadratic equation by a differential operator. He conjectures that this should have one set of solutions yielding the quantum states of the *atom* as such and another set yielding those of the *neutron*. Also he has an argument from which he infers that if the two masses carry electric charges, these would have to be equal and opposite.

Nowadays probably most physicists believe that they ought to be able to do something about the constants of Nature – perhaps to discover relations between them and so to reduce the number of 'independent' constants – perhaps to relate the behaviour of the whole cosmos to the existence and values of the constants – perhaps to discover that some of the constants are not constant (but were it possible to assert that some particular 'constant' varies in a particular way then the coefficients in the formula describing such behaviour would presumably have to be constant) – perhaps even to predict the existence and values of all such constants. At the present time discussions of such matters tend to be linked with consideration of the anthropic principle in its various forms.

Most physicists may not themselves do much about all this, but they seem now to regard it as natural and proper that some physicists should be concerned. If this is so then I think that much of the attitude is due to Eddington's influence. In his own way this aspect of his work did not have a widespread sympathetic reception. Since then his two larger monographs remain impressive but largely incomprehensible. But in spite of this lack of considerable direct effect, it is some measure of Eddington's greatness that the intention of this work seems to have got through to so many physicists. They may not be aware of Eddington as the source of such influence, but their subject has never been the same since he came on the scene.

The Expanding Universe is an almost light-hearted bare introduction to the ideas that Eddington was setting out to develop at the time of writing. It reveals something of the direction of his thinking through the exercise of his rare artistry in presenting it.

1986   William McCrea

Masters, I am to discourse wonders: but ask me not what; for if I tell you, I am no true Athenian. I will tell you everything, right as it fell out.

A Midsummer-Night's Dream

THE EXPANDING UNIVERSE

Chapter I

THE RECESSION OF THE GALAXIES

Pricked out with less and greater lights, between the poles of the universe, the Milky Way so gleameth white as to set very sages questioning.   Dante, *Paradiso*

I

THE first hint of an "expanding universe" is contained in a paper published in November 1917 by Prof. W. de Sitter. Einstein's general theory of relativity had been published two years before, but it had not yet attained notoriety; it was not until the eclipse expeditions of 1919 obtained confirmation of its prediction of the bending of light that public interest was aroused. Meanwhile many investigators had been examining the various consequences of the new theory. Prominent among them was de Sitter who was interested especially in the astronomical consequences. In the course of a highly technical discussion he found that the relativity theory led to an expectation that *the most remote celestial objects would be moving away from us*, or at least that they would deceive the observer into thinking that they were moving away.

De Sitter was perhaps a tipster rather than a prophet. He would not promise anything definitely; but he suggested that we ought to keep a look out for the recession as a rather likely phenomenon. Theory was at the cross-roads, and desired guidance from observation as to which of two possible courses should be pursued. If astronomers were to find a general motion of recession of the most distant objects visible, it would be a strong indication that the road rather fancied by de Sitter was the one to follow. If not, the inference was more doubtful; it might mean that the other road should be followed, or it might only mean that our astronomical survey had not yet been extended to sufficient distance.

Subsequent researches in the field opened up by de Sitter's pioneer investigation have developed and modified his theory. A new point of view has been discovered which renders the results less paradoxical than they appeared originally. We are still led to expect a recession of remote objects, though the recession now predicted is not the original de Sitter effect, which has turned out to be of minor importance. It varies with the distance according to a different law. Moreover, it is a genuine receding motion of remote objects, whereas the phenomenon predicted by de Sitter might be regarded as an imitation recession, and generally was so regarded.

We shall put aside theory for the present, and consider first what astronomical observation tells us. Practically all that I have to relate has been discovered since de Sitter's forecast, much of it within the last four years. These observational results are in some ways so disturbing that there is a natural hesitation in accepting them at their face value. But they have not come upon us like a bolt from the blue, since theorists for the last fifteen years have been half expecting that a study of the most remote objects of the universe might yield a rather sensational development.

The spiral nebulae are the most remote objects known. Rough measurements of their distances have been made, and we place them from 1 million to 150 million light years away; they doubtless extend far beyond the latter distance, but at present it is the limit of our survey. The name "nebula" is applied to different classes of astronomical objects which have nothing in common except a cloudy appearance. There are *gaseous nebulae*, shown by their spectrum to be extremely rarefied gas, either attached to and controlled by a single star or spreading irregularly through a region containing many stars; these are not particularly remote. The *spiral nebulae* on the other hand are extra-galactic objects; that is to say, they lie beyond the limits of the Milky Way aggregation of stars which is the system to which our sun belongs, and are separated from it by wide gulfs of empty space. When we have taken together the sun and all the naked-eye stars and many hundreds of millions of telescopic stars, we have not reached the end of things; we have explored only one island—one oasis in the desert of space. Other islands lie beyond. It is possible with the naked eye to make out a hazy patch of light in the constellation Andromeda which is one of the other islands. A telescope shows many more—an archipelago of island galaxies stretching away one behind another until our sight fails. It is these island galaxies which appear to us as spiral nebulae.

Each island system is believed to be an aggregation of thousands of millions of stars with a general resemblance to our own Milky Way system. As in our own system there may be along with the stars great tracts of nebulosity, sometimes luminous, sometimes dark and obscuring. Many of the nearest systems are seen to have a beautiful double-spiral form (see Frontispiece); and it is believed that the coils of the Milky Way would give the same spiral appearance to our own system if it were viewed from outside. The term "spiral nebula" is, however, to be regarded as a name rather than a description, for it is generally applied to all external galaxies whether they show traces of spiral structure or not.

The island systems are exceedingly numerous. From sample counts it is estimated that more than a million of them are within reach of our present telescopes. If the theory treated in this book is to be trusted, the total number of them must be of the order 100,000,000,000.

In order to fix in our minds the vastness of the system that we shall have to consider, I will give you a "celestial multiplication table". We start with a star as the unit most familiar to us, a globe comparable to the sun. Then—

> A hundred thousand million Stars make one Galaxy;
>
> A hundred thousand million Galaxies make one Universe.

These figures may not be very trustworthy, but I think they give a correct impression.

The lesson of humility has so often been brought home to us in astronomy that we almost automatically adopt the view that our own galaxy is not specially distinguished—not more important in the scheme of nature than the millions of other island galaxies. But astronomical observation scarcely seems to bear this out. According to the present measurements the spiral nebulae, though bearing a general resemblance to our Milky Way system, are distinctly smaller. It has been said that if the spiral nebulae are islands, our own galaxy is a continent. I suppose that my humility has become a middle-class pride, for I rather dislike the imputation that we belong to the aristocracy of the universe. The earth is a middle-class planet, not a giant like Jupiter, nor yet one of the smaller vermin like the minor planets. The sun is a middling sort of star, not a giant like Capella but well above the lowest classes. So it seems wrong that we should happen to belong to an altogether exceptional galaxy. Frankly I do not believe it; it would be too much of a coincidence. I think that this relation of the Milky Way to the other galaxies is a subject on which more light will be thrown by further observational research, and that ultimately we shall find that there are many galaxies of a size equal to and surpassing our own. Meanwhile the question does not much affect the present discussion. If we are in a privileged position, we shall not presume upon it.

I promised to leave aside theory for the present, but I must revert to it for a moment to try to focus our conception of this super-system of galaxies. It is a vista not only of space but of time. A faint cluster of nebulae in Gemini, which at present marks the limit of our soundings of space, takes us back 150 million years into the past—to the time when the light now reaching us started on its journey across the gulf of space. Thus we can scarcely isolate the thought of vast extension from the thought of time and change; and the problem of form and organisation becomes merged in the problem of origin and development. We must, I suppose, imagine the island galaxies to have been formed by gradual condensation of primordial matter. Perhaps in the first stage only the rudiments of matter existed—protons and electrons traversing the void— and the evolution of the elements has progressed simultaneously with the evolution of worlds. Slight condensations occurring here and there by accident would by their gravitating power draw more particles to themselves. Some would quickly disperse again, but some would become firmly established—

> Champions fierce,
> Strive here for mastery, and to battle bring
> Their embryon atoms....To whom these most adhere,
> He rules a moment: Chaos umpire sits,
> And by decision more embroils the fray
> By which he reigns: next him, high arbiter,
> Chance governs all.*

By such conflict the matter of the universe would slowly be collected into islands, leaving comparatively empty spaces from which it had been drained away. We think that one of these original islands has become our Milky Way system, having subdivided again and again into millions of stars. The other islands similarly developed into galaxies, which we see to-day shining as spiral nebulae. It is to these prime units of sub- division of the material universe that our discussion here will relate.

* *Paradise Lost*, Book ii.

If a spiral nebula is not too faint it is possible to measure its radial velocity in the line of sight by measuring the shift of the lines in its spectrum. A valuable early series of such determinations was made by Prof. V. M. Slipher at the Lowell Observatory.

More recently the distances of some of the spiral nebulae have been determined by a fairly trustworthy method. In the nearest spirals it is possible to make out some of the individual stars; but only the most luminous stars, some hundreds or thousands of times brighter than the sun, can be seen at so great a distance. Fortunately among the very brightest of the stars there is a particularly useful class called the Cepheid variables. They vary periodically in brightness owing to an actual pulsation or physical change of the star, the period being anything from a few hours to a few weeks. It has been ascertained from observational study that Cepheids which have the same period are nearly alike in their other properties—luminosity, radius, spectral type, etc. The period is thus a badge, easily recognisable at a distance, which labels the star as having a particular luminosity. For example, if the star is seen to have a period of 10 days, we immediately recognise it as a star of luminosity 950 times greater than the sun. Knowing then its real brightness we put the question, How far off must it be situated so as to be reduced to the faint point of light which we see? The answer gives the distance of the star and of the galaxy in which it lies. This method uses the Cepheid variables as standard candles. If you see a standard candle anywhere and note how bright it appears to you, you can calculate how far off it is; in the same way an astronomer observes his "standard candle" in the midst of a nebula, notes its apparent brightness or magnitude, and deduces the distance of the nebula.

Dr E. P. Hubble at Mount Wilson Observatory was able to discover Cepheid variables in two or three of the nearest spiral nebulae, and so obtained the first real measurement of their distances. Unfortunately this method is not available for the more distant galaxies, and he has had to use more indirect devices for extending the survey. I think that, apart from those distances actually determined by the Cepheid method, we must regard the distances assigned to the spiral nebulae as rather risky estimates; but there is reason to believe that they cannot be entirely misleading, and we shall provisionally accept them here.

When the collected data as to radial velocities and distances are examined a very interesting feature is revealed. The velocities are large, generally very much larger than ordinary stellar velocities. The more distant nebulae have the bigger velocities; the results seem to agree very well with a linear law of increase, the velocity being simply proportional to the distance. The most striking feature is that the galaxies are almost unanimously running away from us.

Let us consider especially the last result and state the observational evidence in more detail. The light of the spiral nebulae, being compounded of the light of a great variety of stars, does not give a good spectrum for measurement. For this reason and because of its faintness the deduced velocities are inaccurate as judged by ordinary standards; but, except for the nearest nebulae, the velocities are themselves so enormous that the error of measurement is comparatively unimportant. Taking the results as published, the present position is that line-of-sight velocities of about 90 galaxies have been measured, and of these only five are moving towards us. At first sight it may seem wrong to pass over the minority as insignificant. But the five exceptions are confined to the very nearest of the nebulae, and their approaching velocities are not large. Since the phenomenon is one which depends on distance (the effect increasing with distance), it is natural that we should have to go out to a fair distance before we find it strong enough to prevail over all other effects (including observational error) so as to display itself uniformly. The five approaching velocities are at least partly attributable to the use of an inappropriate standard of reference. Line-of-sight velocities as published are relative to the sun; but it would be more satisfactory to discuss the velocities relative to our Milky Way system as a whole. It has been found that the sun is pursuing an orbit round the centre of the Milky Way system and has an orbital speed from 200 to 300 kilometres per second. When we correct for this so as to obtain the velocities referred to our galaxy as a whole, the approaching velocities are reduced or disappear. I think it will turn out ultimately that, after all corrections are applied, these nearest nebulae have small receding velocities; for the existence of even one genuine exception would be difficult to explain.

In saying that the speeds of the nebulae are large,

the velocities of ordinary stars are our standard of comparison. For stars in our neighbourhood the individual speed averages 10 to 50 km. per sec. If the speed exceeds 100 km. per sec. the star is described as a "runaway". (We do not here include the above-mentioned orbital motion about the centre of the galaxy which is shared by all stars in the neighbourhood of the sun.) Slipher's first determination of the radial velocities of 40 nebulae included a dozen with velocities from 800–1800 km. per sec. The survey has since been extended to fainter and more distant nebulae by M. L. Humason at Mount Wilson Observatory, and much higher velocities have been found. The speed record is continually being broken. The present holder of the trophy is a nebula forming one of a faint cluster in the constellation Gemini, which is receding with a velocity of 25,000 km. per sec. (15,000 miles per second). This is about the speed of an Alpha particle. Its distance is estimated at 150,000,000 light-years. Doubtless a faster and more distant nebula will have been announced by the time these words are in print.

The simple proportionality of speed to distance was first found by Hubble in 1929. This law is also predicted by relativity theory. According to the original investigation of de Sitter a velocity proportional to the square of the distance would have been expected; but the theory had become better understood since then, and it was already known (though perhaps only to a few*) that simple proportionality to the distance was the correct theoretical result.

* I was not myself aware of it in 1929. For the nature of the change, see p. 49.

According to Hubble's most recent determination, the speed of recession amounts to 550 km. per sec. per megaparsec.* That is to say, a nebula at 1 megaparsec distance should have a speed 550 km. per sec.; at 10 megaparsecs distance, 5500 km. per sec.; and so on. It has been claimed that this determination is accurate to 20 per cent., but I do not think many astronomers take so optimistic a view. The uncertainty lies almost entirely in the scale of nebular distances; there are weak links in the long chain of connection between these vast distances and our terrestrial standard metre. Corrections which have been suggested mostly tend to increase the result; and perhaps the fairest statement is that the velocity of recession is probably between 500 and 1000 km. per sec. per mp.

Specimens of the spectra from which these radial velocities are obtained are shown in Plate II. In the lower four photographs the spectra of the nebulae are the torpedo-shaped black patches; they have terrestrial comparison spectra above and below, which are used to place them in correct vertical alignment. Practically the only recognisable features in the nebular spectra are the H and K lines—two interruptions in the tail of the torpedo where it is fading away. It will be seen that these interruptions move to the right, i.e. to the red end of the spectrum, as we go down the plate. It is this displacement which is measured and gives the receding velocities stated at the foot of the plate.

* 1 megaparsec = 3·26 million light-years.

III

We can exclude the spiral nebulae which are more or less hesitating as to whether they shall leave us by drawing a sphere of rather more than a million light-years radius round our galaxy. *In the region beyond, more than 80 have been observed to be moving outwards, and not one has been found coming in to take their place.*

The inference is that in the course of time all the spiral nebulae will withdraw to a greater distance, evacuating the part of space that we now survey. Ultimately they will be out of reach of our telescopes unless telescopic power is increased to correspond. I find that the observer of nebulae will have to double the aperture of his telescope every 1300 million years merely to keep up with their recession. If we have been thinking that the human race has still billions of years before it in which to find out all that can be found out about the universe, we must count the problem of the spiral nebulae as one of urgency. Let us make haste to study them before they disappear into the distance !

The unanimity with which the galaxies are running away looks almost as though they had a pointed aversion to us. We wonder why we should be shunned as though our system were a plague spot in the universe. But that is too hasty an inference, and there is really no reason to think that the animus is especially directed against our galaxy. If this lecture-room were to expand to twice its present size, the seats all separating from each other in proportion, you would notice that everyone had moved away from you. Your neighbour who was 2 feet away is now 4 feet away; the man over yonder who was 40 feet away is now 80 feet away. It is not *you* they are avoiding; everyone is having the same experience. In a general dispersal or expansion every individual observes every other individual to be moving away from him. The law of a general uniform expansion is that each individual recedes from you at a rate proportional to his distance from you—precisely the law which we observe in the receding motions of the spiral nebulae.*

We shall therefore no longer regard the phenomenon as a movement away from our galaxy. It is a general scattering apart, having no particular centre of dispersal.

I do not wish to insist on these observational facts dogmatically. It is granted that there is a possibility of error and misinterpretation. The survey is just beginning, and things may appear in a different light as it proceeds. But if you ask what is the picture of the universe now in the minds of those who have been engaged in practical exploration of its large-scale features—men not likely to be moved overmuch by ideas of bending of space or the gauge-invariance of the Riemann-Christoffel tensor—I have given you their answer. Their picture is the picture of an *expanding universe*. The super-system of the galaxies is dispersing as a puff of smoke disperses. Sometimes I wonder whether there may not be a greater scale of existence of things, in which it *is* no more than a puff of smoke.

* Our observations determine the *relative* velocity of recession of a nebula, i.e. the rate at which its distance from us is increasing. They do not indicate whether the nebula is moving away from us or we are moving away from the nebula.

For the present I make no reference to any "expansion of space". I am speaking of nothing more recondite than the expansion or dispersal of a material system. Except for the large scale of the phenomenon the expansion of the universe is as commonplace as the expansion of a gas. But nevertheless it gives very serious food for thought. It is perhaps in keeping with the universal change we see around us that time should set a term even to the greatest system of all; but what is startling is the rate at which it is found to be melting away. We do not look for immutability, but we had certainly expected to find a permanence greater than that of terrestrial conditions. But it would almost seem that the earth alters less rapidly than the heavens. The galaxies separate to double their original distances in 1300 million years. That is only of the order of geological time; it is approximately the age assigned to the older rocks in the earth's crust. This is a rude awakening from our dream of leisured evolution through billions of years.*

Such a conclusion is not to be accepted lightly; and those who have cast about for some other interpretation of what seems to have been observed have displayed no more than a proper caution. If the apparent recession of the spiral nebulae is treated as an isolated discovery it is too slender a thread on which to hang far-reaching conclusions; we can only state the bare results of observation, contemplate without much conviction the amazing possibility they suggest, and await further information on the subject.

* I may remind American readers that the English billion is a million million.

If that is not my own attitude, it is because the motion of the remote nebulae does not present itself to me as an isolated discovery. Following de Sitter, I have for fifteen years been awaiting these observational results to see how far they would fall into line with and help to develop the physical theory, which though at first merely suggestive has become much more cogent in the intervening years. After Prof. Weyl's famous extension of the relativity theory I became convinced that the scale of structure of atoms and electrons is determined by the same physical agent that was concerned in de Sitter's prediction. So that hope of progress of a really fundamental kind in our understanding of electrons, protons and quanta is bound up with this investigation of the motions of remote galaxies. Therefore when Dr Hubble hands over a key which he has picked up in intergalactic space, I am not among those who are turning it over and over unable to decide from the look of it whether it is good metal or base metal. The question for me is, Will it unlock the door?

If the observed radial velocities are accepted as genuine, there is no evading the conclusion that the nebulae are rapidly dispersing. The velocities are direct evidence of a hustle which (according to the usual ideas of the rate of evolutionary change) is out of keeping with the character of our staid old universe. Thus the only way of avoiding a great upset of ideas would be to explain away these radial velocities as spurious. What is actually observed is a shifting of the spectrum of the nebula towards the red. Such a shift is commonly caused by the Doppler effect of a receding velocity, in the same way that the pitch of a receding whistle is lowered; but other causes are imaginable. The reddening signifies lower frequency of the light-waves and (in accordance with quantum theory) lower energy; so that if for any cause a light-quantum loses some of its energy in travelling to reach us, the reddening is accounted for without assuming any velocity of the source. For example, the light coming to us from an atom on the sun uses up some of its energy in escaping from the sun's gravitational attraction, and consequently becomes slightly reddened as compared with the light of a terrestrial atom which does not suffer this loss; this is the well-known red shift predicted by Einstein.

In one respect this hypothesis of the loss of energy of nebular light is attractive. If the loss occurs during the passage of the light from the nebula to the observer, we should expect it to be proportional to the distance; thus the red-shift, misinterpreted as a velocity, should be proportional to the distance—which is the law actually found. But on the other hand there is nothing in the existing theory of light (wave theory or quantum theory) which justifies the assumption of such a loss. We cannot without undue dogmatism exclude the possibility of modifications of the existing theory. Light is a queer thing—queerer than we imagined twenty years ago—but I should be surprised if it is as queer as all that.

A theory put forward by Dr Zwicky, that light, by its gravitational effects, parts with its energy to the material particles thinly strewn in intergalactic space which it passes on its way, at one time attracted atten-

Humason

SPECTRA OF NEBULAE

showing lines shifted to the red (to the right), interpreted as
velocity of recession. (See p. 11.)

(1) Sky; velocity, nil. (2) N.G.C. 221; velocity, – 185 km. per sec.
(3) N.G.C. 385; velocity, +4900 km. per sec. (4) N.G.C.
4884; velocity, +6700 km. per sec. (5) Nebula in Leo;
velocity, +19,700 km. per sec.

tion. But the numerical accordance alleged to support his theory turned out to be fallacious, and the suggestion seems definitely untenable.

I think then we have no excuse for doubting the genuineness of the observed velocities—except in so far as they share the general uncertainty that surrounds all our attempts to probe into the secrets of nature.

IV

Now let us turn to theory.

A scientist commonly professes to base his beliefs on observations, not theories. Theories, it is said, are useful in suggesting new ideas and new lines of investigation for the experimenter; but "hard facts" are the only proper ground for conclusion. I have never come across anyone who carries this profession into practice—certainly not the hard-headed experimentalist, who is the more swayed by his theories because he is less accustomed to scrutinise them. Observation is not sufficient. We do not believe our eyes unless we are first convinced that what they appear to tell us is credible.

It is better to admit frankly that theory has, and is entitled to have, an important share in determining belief. For the reader resolved to eschew theory and admit only definite observational facts, *all* astronomical books are banned. *There are no purely observational facts about the heavenly bodies.* Astronomical measurements are, without exception, measurements of phenomena occurring in a terrestrial observatory or station; it is only by theory that they are translated into knowledge of a universe outside.

When an observer reports that he has discovered a new star in a certain position, he is probably unaware that he is going beyond the simple facts of observation. But he does not intend his announcement to be taken as a description of certain phenomena that have occurred in his observatory; he means that he has located a celestial body in a definite direction in interstellar space. He looks on the location as an observational fact—on a surer footing therefore than theoretical inferences such as have been deduced from Einstein's theory. We must break it to him that his supposed "fact", far from being purely observational, is actually an inference based on Einstein's theory—unless, indeed, he has based it on some earlier theory which is even more divorced from observational facts. The observer has given a theoretical interpretation to his measurements by assuming for theoretical reasons that light travels through interstellar space approximately in a straight line. Perhaps he will reply that, in assuming the rectilinear propagation of light, he is not concerned with any theory but is using a fact established by direct experiment. That begs the question how far an experiment under terrestrial conditions can be extrapolated to apply to interstellar space. Surely a reasoned theory is preferable to blind extrapolation. But indeed the observer is utterly mistaken in supposing that the straightness of rays of light assumed in astronomy has been verified by terrestrial experiment. If the rays in interstellar space were no straighter than they are on the earth,* the direction in which a star is seen would be no guide to its actual

* They are deflected by the earth's gravitational field.

position. Light would in fact curl round and come back again before traversing the distance to the nearest star.

Our warrant for concluding that the celestial body is nearly in the direction in which it is seen, is Einstein's theory, which determines the deviation of light from a straight line. Coupled with other theoretical deductions as to the density of matter in interstellar space, it allows us to conclude that the deviation in this case is inappreciable. So if we are willing to use both fact and theory as a basis for belief, we can accept the observer's announcement; but it is not a "hard fact of observation". Although it is a minor point, we may also insist that the theory concerned is Einstein's theory. There was an earlier theory according to which light in empty space travels in straight lines in all circumstances; but since this has been found experimentally to be untrue, it can scarcely be the basis of our observer's conclusion. Perhaps, however, the observer is one of those who do not credit the eclipse observations of the deflection of light, or who deem them insufficient ground for quitting the old theory. If so, he illustrates my dictum that with the hard-headed experimentalist the basis of belief is often theory rather than observation.

My point is that in astronomy it is not a question of whether we are to rely on observation or on theory. The so-called facts are in any case theoretical interpretations of the observations. The only question is, Shall we for this interpretation use the fullest resources of modern theory? For my own part I can see no more reason for preferring the theories of fifty years ago than for preferring the observational data of fifty years ago.

In turning now to the more theoretical side of the problem of the expanding universe, I do not think that we should feel that we are stepping from solid ground into insecurity. Perhaps we are a little safer, for we no longer depend on the interpretation of one type of observation; and our theory comes from the welding together of different lines of physical research. I do not, however, promise security. An explorer is jealous of his reputation for proper caution, but he can never aspire to the quintessence of caution displayed by the man who entrenches himself at home.

V

In 1915 Einstein had by his general theory of relativity brought a large section of the domain of physics into good order. The theory covered *field-physics*, which includes the treatment of matter, electricity, radiation, energy, etc., on the ordinary macroscopic scale perceptible to our senses, but not the phenomena arising from the minute subdivision into atoms, electrons, quanta. For the study of microscopic structure another great theory was being developed—the quantum theory. At that time it lagged far behind, and even now it has not reached the clearness and logical perfection of the relativity theory. It is recognised that the two theories will meet, and that they must ultimately coalesce into one comprehensive theory. The first bridge between them was made by Prof. P. A. M. Dirac in 1928 by his relativity wave-equation of an electron. I hope to show in the last chapter that the recession of the spiral nebulae leads us to the border-land territory between the two theories, where a number of interesting problems await solution. At present, however, we are concerned only with its relation to the theory of relativity.

The central result of Einstein's theory was his law of gravitation, generally expressed in the form $G_{\mu\nu} = 0$, which has the merit of brevity if not of lucidity. We naturally hear most about those rare phenomena in which Einstein's law gives results appreciably different from Newton's law; but it is to be remembered that for ordinary practical purposes the two laws come to the same thing. So if we take $G_{\mu\nu} = 0$ to be the law governing the motions of the spiral nebulae, that is as much as to say they exert the ordinary Newtonian attraction on one another varying as the inverse square of their distance apart. The law throws no light on why the nebulae are running away from us and from one another. The tendency would rather be for the whole system to fall together—though this tendency to collapse might be counteracted as it is in the solar system, for example.

A year or so later Einstein made a slight amendment to his law to meet certain difficulties that he encountered in his theory. There was just one place where the theory did not seem to work properly, and that was—infinity. I think Einstein showed his greatness in the simple and drastic way in which he disposed of difficulties at infinity. He abolished infinity. He slightly altered his equations so as to make space at great distances bend round until it closed up. So that, if in Einstein's space you keep going right on in one direction, you do not get to infinity; you find yourself back at your starting-point again. Since there was no longer any infinity, there could be no difficulties at infinity. Q.E.D.

However, at present we are not concerned with this new kind of space. I only mention it here because I want to speak of the alteration that Einstein made in his law of gravitation. The amended law is written $G_{\mu\nu} = \lambda g_{\mu\nu}$, and contains a natural constant λ called the *cosmical constant*. The term $\lambda g_{\mu\nu}$ is called the *cosmical term*. The constant is so small that in ordinary applications to the solar system, etc., we set it equal to zero, and so revert to the original law $G_{\mu\nu} = 0$. But however small λ may be, the amended law presents the phenomenon of gravitation to us in a new light, and has greatly helped to an understanding of its real significance; moreover, we have now reason to think that λ is not so small as to be entirely beyond observation. The nature of the alteration can be stated as follows: the original law stated that a certain geometrical characteristic $(G_{\mu\nu})$ of empty space is always zero; the revised law states that it is always in a constant ratio to another geometrical characteristic $(g_{\mu\nu})$. We may say that the first form of the law utterly dissociates the two characteristics by making one of them zero and therefore independent of the other; the second form intimately connects them. Geometers can invent spaces which have not either of these properties; but actual space, surveyed by physical measurement, is not of so unlimited a nature.

We have already said that the original term in the law gives rise to what is practically the Newtonian attraction between material objects. It is found similarly that the added term $(\lambda g_{\mu\nu})$ gives rise to a repulsion directly proportional to the distance. Distance from what? Distance from *anywhere*; in particular, distance from the observer. It is a dispersive force like that which I imagined as scattering apart the audience in the lecture-room. Each thinks it is directed away from him. We may say that the repulsion has no centre, or that every point is a centre of repulsion.

Thus in straightening out his law of gravitation to satisfy certain ideal conditions, Einstein almost inadvertently added a repulsive scattering force to the Newtonian attraction of bodies. We call this force the *cosmical repulsion*, for it depends on and is proportional to the cosmical constant. It is utterly imperceptible within the solar system or between the sun and neighbouring stars. But since it increases proportionately to the distance we have only to go far enough to find it appreciable, then strong, and ultimately overwhelming. In practical observation the farthest we have yet gone is 150 million light-years. Well within that distance we find that celestial objects are scattering apart as if under a dispersive force. Provisionally we conclude that here cosmical repulsion has become dominant and is responsible for the dispersion.

We have no *direct* evidence of an outward acceleration of the nebulae, since it is only the velocities that we observe. But it is reasonable to suppose that the nebulae, individually as well as collectively, follow the rule—the greater the distance the faster the recession. If so, the velocity increases as the nebula recedes, so that there is an outward acceleration. Thus from the observed motions we can work backwards and calculate the repulsive force, and so determine observationally the cosmical constant λ.

Much turns on whether Einstein was really justified in making the change in his law of gravitation which introduced this cosmical repulsion. His original reason was not very convincing, and for some years the cosmical term was looked on as a fancy addition rather than as an integral part of the theory. Einstein has been as severe a critic of his own suggestion as anyone, and he has not invariably adhered to it. But the cosmical constant has now a secure position owing to a great advance made by Prof. Weyl, in whose theory it plays an essential part.* Not only does it unify the gravitational and electromagnetic fields, but it renders the theory of gravitation and its relation to space-time measurement so much more illuminating, and indeed self-evident, that return to the earlier view is unthinkable. I would as soon think of reverting to Newtonian theory as of dropping the cosmical constant.

VI

Let us now review the position. According to relativity theory the complete field of force contains besides the ordinary Newtonian attraction a repulsive (scattering) force varying directly as the distance. It is well known that Einstein's law differs slightly from Newton's, giving for example an extra effect which

---

* "The cosmological factor which Einstein added to his theory later is part of ours from the very beginning." *Raum. Zeit. Materie*, p. 297 (English Edition).

has been detected in the orbit of the fast-moving planet Mercury; the cosmical repulsion is another point of difference between them, detectable only in the motions of remote objects. From a theoretical standpoint I think there is no more doubt about the cosmical repulsion than about the force which perturbs Mercury; but it does not admit of so decisive an observational test. As regards Mercury the theoretical prediction was quantitative; but relativity theory does not indicate any particular magnitude for the cosmical repulsion. A merely qualitative test is never very conclusive.

However, so far as it goes, the test is satisfactory. We do find observationally a dispersion of the system of the galaxies such as would be caused by the predicted repulsion. The motions are extremely large and the effect stands out clearly above all minor irregularities. The theory thus clears its first hurdle with some *éclat*; whether it will win the race is another question. Although the test is not quantitative it is more far-reaching than is sometimes supposed. There are only two ways of accounting for large receding velocities of the nebulae: (1) they have been produced by an outward directed force as we here suppose, or (2) as large or larger velocities have existed from the beginning of the present order of things.* Several

* For completeness we must add the possible hypothesis that the system once extended much further than now, that it collapsed, and is now on the rebound. This allows the large velocities to have been produced by *inward* directed force, the inward velocities being turned into outward velocities by passage through the centre. So far as I know, this is not advocated by anyone. It does not seem capable of providing for the distribution of velocities which we observe.

rival explanations of the recession of the nebulae, which do not accept it as evidence of repulsive force, have been put forward. These necessarily adopt the second alternative, and postulate that the large velocities have existed from the beginning. This might be true; but it can scarcely be called an *explanation* of the large velocities.

Our best hope of further progress is to discover some additional test for the theory—if possible, a stringent quantitative test. We want to predict the actual magnitude of the cosmical repulsion, and see if the observed motions of the nebulae confirm the predicted value. Relativity theory alone cannot do this, but when relativity is combined with wave-mechanics the quantitative prediction seems possible. This development is explained in Chapter IV.

Thus far we have been treating a fairly straightforward subject. Apart from the vast magnitudes involved there is nothing that particularly taxes the imagination. In the next chapter I shall present a rather different view involving difficult conceptions. I can imagine the reader saying, "Why do you spoil it all, just when I was beginning to see what it is all about?"

If I introduce a different kind of outlook it is because I am going on to treat of regions of the universe beyond those that we have hitherto considered. Primarily the present chapter deals with the region actually explored, up to 150 million light-years' distance. If the galaxies come to an end there, no more need be said; the points discussed in the next chapter are scarcely relevant and its outlook is unnecessarily pedantic. But there is no sign that the system of the galaxies is coming to an end, and presumably it extends considerably beyond 150 million light-years. It might extend to, say, five times that distance without any important new feature; but if we have to go much beyond that, there is trouble in store. The appropriate speed of recession would be beginning to approach the velocity of light—a point which evidently requires looking into. We have a force of cosmical repulsion, increasing with the distance, which is already rather powerful; if we go on to a vastly greater distance something must give way at last—only Einstein has taken the precaution of closing up the universe to prevent us from going too far.

The object of the ensuing development is to deal with questions which arise as to the possible extension of the system of the galaxies beyond the region at present explored. We shall consider extrapolation in time as well as in space, and discuss the history of evolution of the system.

What is the object of making these risky extrapolations to regions of space and time remote from our practical experience? It might be a sufficient answer to say that we are *explorers*. But there is another and more urgent reason. The man who for the first time sees an aeroplane passing overhead doubtless wonders how it goes. I do not think he can be accused of eccentricity if he also wonders *how it stops*. It is true that he sees no signs of its stopping; he is mentally extrapolating the flight beyond the range that is visible. He cannot be sure of his extrapolation; outside his range of vision there may be conditions, of which he is unaware, which will stop the flight in a manner different from his conjecture. But he will have much more confidence in his conclusions as to the mechanism of the aeroplane if they will explain the flight from start to stop without postulating some unknown intervention. At first sight it seems a reasonable programme for science to tidy up the region of space and time of which we have some experience and not to theorise about what lies beyond; but the danger of such a limitation is that the tidying up may consist in taking the difficulties and inexplicabilities and dumping them over the border instead of really straightening them out.

We have seen that there is a force of cosmical repulsion growing larger as the distance from us increases. At the greatest distance yet explored it is still increasing. The foregoing theory explains how it goes. But we have still a desire to understand how it stops.

Chapter II

SPHERICAL SPACE

I could be bounded in a nutshell and count myself a king of infinite
space. *Hamlet*

I

WHEN a physicist refers to curvature of space he at
once falls under suspicion of talking metaphysics. Yet
space is a prominent feature of the physical world;
and measurement of space—lengths, distances, volumes
—is part of the normal occupation of a physicist.
Indeed it is rare to find any quantitative physical
observation which does not ultimately reduce to
measuring distances. Is it surprising that the precise
investigation of physical space should have brought to
light a new property which our crude sensory per-
ception of space has passed over?

Space-curvature is a purely physical characteristic
which we may find in a region by suitable experiments
and measurements, just as we may find a magnetic
field. In curved space the measured distances and
angles fit together in a way different from that with
which we are familiar in the geometry of flat space;
for example, the three angles of a triangle do not add
up to two right angles. It seems rather hard on the
physicist, who conscientiously measures the three
angles of a triangle, that he should be told that if the
sum comes to two right angles his work is sound
physics, but if it differs to the slightest extent he is
straying into metaphysical quagmires.

In using the name "curvature" for this characteristic of space, there is no metaphysical implication.
The nomenclature is that of the pure geometers who
had already imagined and described spaces with this
characteristic before its actual physical occurrence
was suspected.

Primarily, then, curvature is to be regarded as the
technical name for a property discovered observationally. It may be asked, How closely does "curvature" as a technical scientific term correspond to the
familiar meaning of the word? I think the correspondence is about as close as in the case of other
familiar words, such as Work, Energy, Probability,
which have acquired a specialised meaning in science.

We are familiar with curvature of *surfaces*; it is a
property which we can impart by bending and deforming a flat surface. If we imagine an analogous
property to be imparted to *space* (three-dimensional)
by bending and deforming it, we have to picture an
extra dimension or direction in which the space is
bent. There is, however, no suggestion that the extra
dimension is anything but a fictitious construction,
useful for representing the property pictorially, and
thereby showing its mathematical analogy with the
property found in surfaces. The relation of the picture
to the reality may perhaps best be stated as follows.
In nature we come across curved surfaces and curved
spaces, i.e. surfaces and spaces exhibiting the observational property which has been technically called
"curvature". In the case of a surface we can ourselves
remove this property by bending and deforming it;
we can therefore conveniently describe the property

by the operation (bending or curving) which we should have to perform in order to remove it. In the case of a space we cannot ourselves remove the property; we cannot alter space artificially as we alter surfaces. Nevertheless we may conveniently describe the property by the imaginary operation of bending or curving, which would remove it if it could be performed; and in order to use this mode of description a fictitious dimension is introduced which would make the operation possible.

Thus if we are not content to accept curvature as a technical physical characteristic but ask for a picture giving fuller insight, we have to picture more than three dimensions. Indeed it is only in simple and symmetrical conditions that a fourth dimension suffices; and the general picture requires six dimensions (or, when we extend the same ideas from space to space-time, ten dimensions are needed). That is a severe stretch on our powers of conception. But I would say to the reader, do not trouble your head about this picture unduly; it is a stand-by for very occasional use. Normally, when reference is made to space-curvature, picture it as you picture a magnetic field. Probably you do *not* picture a magnetic field; it is something (recognisable by certain tests) which you use in your car or in your wireless apparatus, and all that is needed is a recognised name for it. Just so; space-curvature is something found in nature with which we are beginning to be familiar, recognisable by certain tests, for which ordinarily we need not a picture but a name.

It is sometimes said that the difference between the

mathematician and the non-mathematician is that the former can picture things in four dimensions. I suppose there is a grain of truth in this, for after working for some time in four or more dimensions one does involuntarily begin to picture them after a fashion. But it has to be added that, although the mathematician visualises four dimensions, his picture is *wrong* in essential particulars—at least mine is. I see our spherical universe like a bubble in four dimensions; length, breadth, and thickness, all lie in the skin of the bubble. Can I picture this bubble rotating? Why, of course I can. I fix on one direction in the four dimensions as axis, and I see the other three dimensions whirling round it. Perhaps I never actually see more than two at a time; but thought flits rapidly from one pair to another, so that all three seem to be hard at it. Can *you* picture it like that? If you fail, it is just as well. For we know by analysis that a bubble in four dimensions does not rotate that way at all. Three dimensions cannot spin round a fourth. They must rotate two round two; that is to say, the bubble does not rotate about a line axis but about a plane. I know that that is true; but I cannot visualise it.

I need scarcely say that our scientific conclusions about the curvature of space are not derived from the false involuntary picture, but by algebraic working out of formulae which, though they may be to some extent illustrated by such pictures, are independent of pictures. In fact, the pictorial conception of space-curvature falls between two stools: it is too abstruse to convey much illumination to the non-mathematician, whilst the mathematician practically ignores it and relies on the more dependable and more powerful algebraic methods of investigating this property of physical space.

Having said so much in disparagement of the picture of our three-dimensional space contorted by curvature in fictitious directions, I must now mention one application in which it is helpful. We are assured by analysis that in one important respect the picture is not misleading. The curvature, or bending round of space, may be sufficient to give a "closed space"— space in which it is impossible to go on indefinitely getting farther and farther from the starting-point. Closed space differs from an open infinite space in the same way that the surface of a sphere differs from a plane infinite surface.

II

We may say of the surface of a sphere (1) that it is a *curved* surface, (2) that it is a *closed* surface. Similarly we have to contemplate two possible characteristics of our actual three-dimensional space, *curvature* and *closure*. A closed surface or space must necessarily be curved, but a curved surface or space need not be closed. Thus the idea of closure goes somewhat beyond the idea of curvature; and, for example, it was not contemplated in the first announcement of Einstein's general relativity theory which introduced curved space.

In the ordinary application of Einstein's theory to the solar system and other systems on a similar scale the curvature is small and amounts only to a very slight wrinkling or hummocking. The distortion is local, and does not affect the general character of space as a whole. Our present subject takes us much farther afield, and we have to apply the theory to the great super-system of the galaxies. The small local distortions now have cumulative effect. The new investigations suggest that the curvature actually leads to a complete bending round and closing up of space, so that it becomes a domain of finite extent. It will be seen that this goes beyond the original proposal; and the evidence for it is by no means so secure. But all new exploration passes through a phase of insecurity.

For the purpose of discussion this closed space is generally taken to be spherical. The presence of matter will cause local unevenness; the scale that we are now contemplating is so vast that we scarcely notice the stars, but the galaxies change the curvature locally* and so pull the sphere rather out of shape. The ideal spherical space may be compared to the geoid used to represent the average figure of the earth with the mountains and ocean beds smoothed away. It may be, however, that the irregularity is much greater, and the universe may be pear-shaped or sausage-shaped; the 150 million light-years over which our observational survey extends is only a small fraction of the whole extent of space, so that we are not in a position to dogmatise as to the actual shape. But we can use the spherical world as a typical model,

---

* Einstein's law of gravitation connects the various components of curvature of space with the density, momentum, and stress of the matter occupying it. I would again remind the reader that space-curvature is the technical name for an observable physical property, so that there is nothing metaphysical in the idea of matter producing curvature any more than in a magnet producing a magnetic field.

which will illustrate the peculiarities arising from the closure of space.

In spherical space, if we go on in the same direction continually, we ultimately reach our starting-point again, having "gone round the world". The same thing happens to a traveller on the earth's surface who keeps straight on bearing neither to the left nor to the right. Thus the closure of space may be thought of as analogous to the closure of a surface, and generally speaking it has the same connection with curvature. The whole area of the earth's surface is finite, and so too the whole volume of spherical space is finite. It is "finite but unbounded"; we never come to a boundary, but owing to the re-entrant property we can never be more than a limited distance away from our starting-point.

In the theory that I am going to describe the galaxies are supposed to be distributed throughout a closed space of this kind. As there is no boundary— no point at which we can enter or leave the closed space—this constitutes a self-contained finite universe.

Perhaps the most elementary characteristic of a spherical universe is that at great distances from us there is not so much room as we should have anticipated. On the earth's surface the area within 2 miles of Charing Cross is very nearly 4 times the area within 1 mile; but at a distance of say 4000 miles this simple progression has broken down badly. Similarly in the universe the volume, or amount of room, within 2 light-years of the sun is very nearly 8 times the volume within 1 light-year; but the volume within 4000 million light-years of the sun is

35

considerably less than 8 times the volume within 2000 million light-years. We have no right to be surprised. How could we have expected to know how much room there would be out there without examining the universe to see? It is a common enough experience that simple rules, which hold well enough for a limited range of trial, break down when pushed too far. There is no juggling with words in these statements; the meaning of distance and volume in surveying the earth or the heavens is not ambiguous; and although there are practical difficulties in measuring these vast distances and volumes there is no uncertainty as to the ideal that is aimed at. I do not suggest that we have checked by direct measurement the falling off of volume at great distances; like many scientific conclusions, it is a very indirect inference. But at least it has been reached by examining the universe; and, however shaky the deduction, it has more weight than a judgment formed without looking at the universe at all.

Much confusion of thought has been caused by the assertion so often made that we can use any kind of space we please (Euclidean or non-Euclidean) for representing physical phenomena, so that it is impossible to disprove Euclidean space observationally. We can graphically represent (or misrepresent) things as we please. It is possible to represent the curved surface of the earth in a flat space as, for example, in maps on Mercator's projection; but this does not render meaningless the labours of geodesists as to the true figure of the earth. Those who *on this ground* defend belief in a flat universe must also defend belief in a flat earth.

III

There is a widespread impression, which has been encouraged by some scientific writers, that the consideration of spherical space in this subject is an unnecessary mystification, and that we could say all we want to say about the expanding system of the galaxies without using any other conception than that of Euclidean infinite space. It is suggested that talk about expanding space is mere metaphysics, and has no real relevance to the expansion of the material universe itself, which is commonplace and easily comprehensible. This is a mistaken idea. The general phenomenon of expansion, including the explanation provided by relativity theory, can be expounded up to a certain point without any recondite conceptions of space, as has been done in Chapter 1; but there are other consequences of the theory which cannot be dealt with so simply. To consider these we have to change the method, and partly transfer our attention from the properties and behaviour of the material system to the properties and behaviour of the space which it occupies. This is necessary because the properties attributed to the material system by the theory are so unusual that they cannot even be described without self-contradiction if we continue to picture the system in flat (i.e. Euclidean) space. This does not constitute an objection to the theory, for there is, of course, no reason for supposing space to be flat unless our observations show it to be flat; and there is no reason why we should be able to picture or describe the system in flat space if it is not in flat space. It is no disparagement to a square peg to say that it will not fit into a round hole.

I will liken the super-system of galaxies (the universe) to a peg which is fitted into a hole—space. In Chapter 1 we were only concerned with a little bit of the peg (the 150 million light-years surveyed) and the question of fit scarcely arose. When we turn to consider the whole peg we find mathematically that, unless something unforeseen occurs beyond the region explored, it is (for the purposes of this analogy) a square peg. Immediately there is an outcry: "That is an impossible sort of peg—not really a peg at all". Our answer is that it is an excellent peg, as good as any on the market, provided that you do not want to fit it into a round hole. "But holes are round. It is the nature of holes to be round. A Greek two thousand years ago said that they are round." And so on. So whether I want it or not, the argument shifts from the peg to the hole—the space into which the material universe is fitted. It is over the hole that the battle has been fought and won; I think now that every authority admits, if only grudgingly, that the square hole—by which I here symbolise closed space—is a physical possibility.

The issue that I am here dealing with is not whether the theory of a closed expanding universe is right or wrong, probable or improbable, but whether, if we hold the theory, spherical space is necessary to the statement of it. I am not here replying to those who disbelieve the theory, but to those who think its strangeness is due to the mystifying language of its exponents. The following will perhaps show that there has been no gratuitous mystification:

I want you to imagine a system of say a billion stars spread approximately uniformly so that each star has neighbours surrounding it on all sides, the distance of each star from its nearest neighbours being approximately the same everywhere. (Lest there be any doubt as to the meaning of *distance*, I define it as the distance found by parallax observation, or by any other astronomical method accepted as equivalent to actually stepping out the distance.) Can you picture this?

— Yes. Except that you forgot to consider that the system will have a boundary; and the stars at the edge will have neighbours on one side only, so that they must be excepted from your condition of having neighbours on all sides.

— No; I meant just what I said. I want *all* the stars to have neighbours surrounding them. If you picture a place where the neighbours are on one side only— what you call a boundary—you are not picturing the system I have in mind.

— But your system is impossible; there must be a boundary.

— Why is it impossible? I could arrange a billion people on the surface of the earth (spread over the whole surface) so that each has neighbours on all sides, and no question of a boundary arises. I only want you to do the same with the stars.

— But that is a distribution over a surface. The stars are to be distributed in three-dimensional space, and space is not like that.

— Then you agree that if space could be "like that" my system would be quite possible and natural?

39

— I suppose so. But how could space be like that?

— We will discuss space if you wish. But just now when I was trying to explain that according to present theory space does behave "like that", I was told that the discussion of space was an unnecessary mystification, and that if I would stick to a description of my material system it would be seen to be quite commonplace and comprehensible. So I duly described my material system; whereupon *you* immediately raised questions as to the nature of space.

In the spherical universe the character of the material system is as peculiar as the character of the space. The material system, like the space, exhibits *closure*; so that no galaxy is more central than another, and none can be said to be at the outside. Such a distribution is at first sight inconceivable, but that is because we try to conceive it in flat space. The space and the material system have to fit one another. It is no use trying to imagine the system of galaxies contemplated in Einstein's and Lemaître's theories of the universe, if the only kind of space in our minds is one in which such a system cannot exist.

In the foregoing conversation I have credited the reader with a feeling which instinctively rejects the possibility of a spherical space or a closed distribution of galaxies. But spherical space does not contradict our experience of space, any more than the sphericity of the earth contradicts the experience of those who have never travelled far enough to notice the curvature. Apart from our reluctance to tackle a difficult and unfamiliar conception, the only thing that can be

urged against spherical space is that more than twenty
centuries ago a certain Greek published a set of
axioms which (inferentially) stated that spherical
space is impossible. He had perhaps more excuse, but
no more reason, for his statement than those who
repeat it to-day.

Few scientific men nowadays would reject spherical
space as impossible, but there are many who take the
attitude that it is an unlikely kind of hypothesis only
to be considered as a last resort. Thus, in support of
some of the proposed explanations of the motions of
the spiral nebulae, it is claimed that they have the
"advantage" of not requiring curved space. But
what is the supposed disadvantage of curved space?
I cannot remember that any disadvantage has ever
been pointed out. On the other hand it is well known
that the assumption of flat physical space leads to very
serious theoretical and logical difficulties, as will be
explained later (p. 102).

A closed system of galaxies requires a closed space.
If such a system expands, it requires an expanding
space. This can be seen at once from the analogy that
we have already used, viz. human beings distributed
evenly over the surface of the earth; clearly they cannot scatter apart from one another unless the earth's
surface expands.

This should make clear how the present theory of
the expanding universe stands in relation to (a) the
expansion of a material system, and (b) the expansion
of space. The observational phenomenon chiefly concerned (recession of the spiral nebulae) is obviously
expansion of a material system; and the onlooker is

often puzzled to find theorists proclaiming the doctrine of an expanding space. He suspects that there has been confusion of thought of a rather elementary kind. Why should not the space be there already, and the material system expand into it, as material systems usually do? If the system of galaxies comes to an end not far beyond the greatest distance we have plumbed, then I agree that that is what happens. But the system shows no sign of coming to an end, and, if it goes on much farther, it will alter its character. This change of character is a matter of mathematical computation which cannot be discussed here; I need only say that it is connected with the fact that, if the speed of recession continues to increase outwards, it will ere long approach the speed of light, so that something must break down. The result is that the system becomes a closed system; and we have seen that such a system cannot expand without the space also expanding. That is how *expansion of space* comes in. I daresay that (for historical reasons) expansion of space has often been given too much prominence in expositions of the subject, and readers have been led to think that it is more directly concerned in the explanation of the motions of the nebulae than is actually the case. But if we are to give a full account of the views to which we are led by theory and observation, we must not omit to mention it.

What I have said has been mainly directed towards removing preliminary prejudices against a closed space or a closed system of galaxies. I do not suggest that the reasons for adopting closed space are overwhelmingly strong;* but even slight advantages may be of weight when there is nothing to place in the opposite scale. If we adopt open space we encounter certain difficulties (not necessarily insuperable) which closed space entirely avoids; and we do not want to divert the inquiry into a speculation as to the solution of difficulties which need never arise. If we wish to be non-committal, we shall naturally work in terms of a closed universe of finite radius R, since we can at any time revert to an infinite universe by making R infinite.

There is one other type of critic to whom a word may be said. He feels that space is not solely a matter that concerns the physicists, and that by their technical definitions and abstractions they are making of it something different from the common man's space. It would be difficult to define precisely what is in his mind. Perhaps he is not thinking especially of space as a measurable constituent of the physical universe, and is imagining a world order transcending the delusions of our sensory organs and the limitations of our micrometers—a space of "things as they really are". It is no part of my present subject to discuss the relation of the world as conceived in physics to a wider interpretation of our experience; I will only say that that part of our conscious experience representable by physical symbols ought not to claim to be the whole. As a conscious being *you* are not one of my symbols; your domain is not circumscribed by my spatial measurements. If, like Hamlet, you count yourself

---

* *Curved* space is fundamental in relativity theory, and the argument for adopting it is generally considered to be overwhelming. It is *closed* space which needs more evidence.

king of an infinite space, I do not challenge your sovereignty. I only invite attention to certain disquieting rumours which have arisen as to the state of Your Majesty's nutshell.

IV

The immediate result of introducing the cosmical term into the law of gravitation was the appearance (in theory) of two universes—the Einstein universe and the de Sitter universe. Both were closed spherical universes; so that a traveller going on and on in the same direction would at last find himself back at the starting-point, having made a circuit of space. Both claimed to be static universes which would remain unchanged for any length of time; thus they provided a permanent framework within which the small scale systems—galaxies and stars—could change and evolve. There were, however, certain points of difference between them. An especially important difference, because it might possibly admit of observational test, was that in de Sitter's universe there would be an apparent recession of remote objects, whereas in Einstein's universe this would not occur. At that time only three radial velocities of spiral nebulae were known, and these somewhat lamely supported de Sitter's universe by a majority of 2 to 1. There the question rested for a time. But in 1922 Prof. V. M. Slipher furnished me with his (then unpublished) measures of 40 spiral nebulae for use in my book *Mathematical Theory of Relativity*. As the majority had become 36 to 4, de Sitter's theory began to appear in a favourable light.

The Einstein and de Sitter universes were two alternatives arising out of the same theoretical basis. To give an analogy—suppose that we are transported to a new star, and that we notice a number of celestial bodies in the neighbourhood. We should know from gravitational theory that their orbits must be either ellipses or hyperbolas; but only observation can decide which. Until the observational test is made there are two alternatives; the objects may have elliptic orbits and constitute a permanent system like the solar system, or they may have hyperbolic orbits and constitute a dispersing system. Actually the question whether the universe would follow Einstein's or de Sitter's model depended on how much matter was present in the universe,—a question which could scarcely be settled by theory—and is none too easy to settle by observation.

We have now realised that the changelessness of de Sitter's universe was a mathematical fiction. Taken literally his formulae described a *completely empty* universe; but that was meant to be interpreted generously as signifying that the average density of matter in it, though not zero, was low enough to be neglected in calculating the forces controlling the system. It turned out, however, that the changelessness depended on there being literally no matter present. In fact the "changeless universe" had been invented by the simple expedient of omitting to put into it anything that could exhibit change. We therefore no longer rank de Sitter's as a static universe; and Einstein's is the only form of material universe which is genuinely static or motionless.

The situation has been summed up in the statement that Einstein's universe contains matter but no motion and de Sitter's contains motion but no matter. It is clear that the actual universe containing both matter and motion does not correspond exactly to either of these abstract models. The only question is, Which is the better choice for a first approximation? Shall we put a little motion into Einstein's world of inert matter, or shall we put a little matter into de Sitter's Primum Mobile?

The choice between Einstein's and de Sitter's models is no longer urgent because we are not now restricted to these two extremes; we have available the whole chain of intermediate solutions between motionless matter and matterless motion, from which we can pick out the solution with the right proportion of matter and motion to correspond with what we observe. These solutions were not sought earlier, because their appropriateness was not realised; it was the preconceived idea that a static solution was a necessity in order that everything might be referred to an unchanging background of space. We have seen that this requirement should strictly have barred out de Sitter's solution, but by a fortunate piece of gate-crashing it gained admission; it was the precursor of the other non-static solutions to which attention is now mainly directed.

The deliberate investigation of non-static solutions was carried out by A. Friedmann in 1922. His solutions were rediscovered in 1927 by Abbé G. Lemaître, who brilliantly developed the astronomical theory resulting therefrom. His work was published in a rather inaccessible journal, and seems to have remained unknown until 1930 when attention was called to it by de Sitter and myself. In the meantime the solutions had been discovered for the third time by H. P. Robertson, and through him their interest was beginning to be realised. The astronomical application, stimulated by Hubble and Humason's observational work on the spiral nebulae, was also being rediscovered, but it had not been carried so far as in Lemaître's paper.

The intermediate solutions of Friedmann and Lemaître are "expanding universes". Both the material system and the closed space, in which it exists, are expanding. At one end we have Einstein's universe with no motion and therefore in equilibrium. Then, as we proceed along the series, we have model universes showing more and more rapid expansion until we reach de Sitter's universe at the other end of the series. The rate of expansion increases all the way along the series and the density diminishes; de Sitter's universe is the limit when the average density of celestial matter approaches zero. The series of expanding universes then stops, not because the expansion becomes too rapid, but because there is nothing left to expand.

We can better understand this series of models by starting at the de Sitter end. As explained in Chapter 1 there are two forces operating, the ordinary Newtonian attraction between the galaxies and the cosmical repulsion. In the de Sitter universe the density of matter is infinitely small so that the Newtonian attraction is negligible. The cosmical repulsion acts without check, and we get the greatest possible rate of expansion of the system. When more matter is inserted, the mutual gravitation tends to hold the mass together and opposes the expansion. The more matter put in, the slower the expansion. There will be a particular density at which the Newtonian attraction between the galaxies is just strong enough to counterbalance the cosmical repulsion, so that the expansion is zero. This is Einstein's universe. If we put in still more matter, attraction outweighs repulsion and we obtain a model of a contracting universe.

Primarily this series of models is a series of alternatives, one of which has to be selected to represent our actual universe. But it has a still more interesting application. As time goes on the actual universe travels along the series of models, so that the whole series gives a picture of its life-history. At the present moment the universe corresponds to a particular model; but since it is expanding its density is diminishing. Therefore a million years hence we shall need a model of lower density, i.e. nearer to the de Sitter end of the series.

Tracing this progression as far back as possible we reach the conclusion that the world started as an Einstein universe; it has passed continuously along the series of models having more and more rapid expansion; and it will finish up as a de Sitter universe.

Allusion has been made to the fact that the recession of the galaxies in the present theory of the expanding universe is not precisely the effect foreseen by de Sitter. It may be well to explain the manner of the transition.

The phenomenon that is generally called the "de Sitter effect" was a rather mysterious slowing down of time at great distances from the observer; atomic vibrations would be executed more slowly, so that their light would be shifted to the red and imitate the effect of a receding velocity. But besides discovering this, de Sitter examined the equations of motion and noticed that the *real* velocities of distant objects would probably be large; he did not, however, expect these real velocities to favour recession rather than approach. I am not sure when it was first recognised that the complication in the equations of motion was neither more nor less than a repulsive force proportional to the distance; but it must have been before 1922. Summarising the theory at that date, I wrote— "De Sitter's theory gives a double explanation of this motion of recession: first, there is the general tendency to scatter according to the equation $d^2r/ds^2 = \frac{1}{3}\lambda r$; second, there is the general displacement of spectral lines to the red in distant objects due to the slowing down of atomic vibrations which would be erroneously interpreted as a motion of recession".* I also pointed out that it was a question of definition whether the latter effect should be regarded as a spurious or a genuine velocity. During the time that its light is travelling to us, the nebula is being accelerated by the cosmical repulsion and acquires an additional outward velocity exceeding the amount in dispute; so that the velocity, which was spurious at the time of emission of the light, has become genuine by the time of its arrival. Inferentially this meant that the slowing

* *Mathematical Theory of Relativity*, p. 161.

down of time had become a very subsidiary effect compared with cosmical repulsion; but this was not so clearly realised as it might have been. The subsequent developments of Friedmann and Lemaître were geometrical and did not allude to anything so crude as "force"; but, examining them to see what has happened, we find that the slowing down of time has been swallowed up in the cosmical repulsion; it was a small portion of the whole effect (a second-order term) which had been artificially detached by the earlier methods of analysis.

V

An Einstein universe is in equilibrium, but its equilibrium is unstable. The Newtonian attraction and the cosmical repulsion are in exact balance. Suppose that a slight disturbance momentarily upsets the balance; let us say that the Newtonian attraction is slightly weakened. Then repulsion has the upper hand, and a slow expansion begins. The expansion increases the average distance apart of the material bodies so that their attraction on one another is lessened. This widens the difference between attraction and repulsion, and the expansion becomes faster. Thus the balance becomes more and more upset until the universe becomes irrevocably launched on its course of expansion. Similarly if the first slight disturbance were a strengthening of the Newtonian attraction, this would cause a small contraction. The material systems would be brought nearer together and their mutual attraction further increased. The contracting tendency thus becomes more and more reinforced. Einstein's universe is delicately poised so that the slightest disturbance will cause it to topple into a state of ever-increasing expansion or of ever-increasing contraction.

The original unstable Einstein universe might have turned into an expanding universe or into a contracting universe. Apparently it has chosen expansion. The question arises, Can we explain this choice? I do not think it will be any grave discredit if we fail, for I cannot recall any other case in which theory has succeeded in predicting which way an unstable body will fall. However we shall try. We have to consider what kind of spontaneous disturbance could occur in the primordial distribution of matter from which our galaxies and stars have been evolved; for definiteness I picture it as a motionless uniform nebula filling the spherical world. Two kinds of spontaneous change have been suggested:

(1) The matter will form local condensations so as to become unevenly distributed.

(2) Material mass may become converted into radiation, either in the process of building up complex atoms (e.g. the formation of helium from hydrogen) or in the mutual annihilation of electrons and protons.

It can be shown that the conversion of material mass into radiation would start a contraction. Mass for mass, radiation is more effective than matter in exerting gravitational attraction; hence the conversion tips the balance in favour of contraction. Accordingly our hopes of explaining the decision to expand must rest on process (1). The investigation is peculiarly difficult, because it turns out that to a first approximation the redistribution of matter in condensations makes no difference to the balance, and it is necessary to carry the calculation to a high approximation to obtain the deciding term. The problem has been treated by McVittie, McCrea, Lemaître and Sen—not always with accordant results; and I doubt whether I am qualified to judge on so technical a question. I am inclined to think, however, that Lemaître's treatment goes to the root of the matter.*

It has been mentioned that, although we often consider models of the universe which are perfectly spherical, the actual universe must be more irregular. A better approximation would be a pimply sphere—the pimples corresponding to the galaxies; for wherever there is matter, the curvature is locally increased. Whilst a "pimply Einstein world" would have approximately the same properties as an ideal Einstein world, it was at first thought that exact equilibrium was only possible for the exact sphere. It is found, however, that a pimply sphere can also be in exact equilibrium and form a static universe. This was pointed out explicitly by Prof. N. K. Sen, who has given a simple and elegant treatment; but it appears to have been implicit in the earlier work of Lemaître.

Suppose for a moment that, when a condensation is formed, the condensation separates completely from the surrounding matter and leaves an empty crack all round. We imagine that a sphere of gas is separated in this way, and continues to condense more and more. Lemaître (by extending a theorem due to Birkhoff)

* Lemaître's paper (*Monthly Notices*, vol. XCI, p. 490) seems to me very obscure, but I have had the advantage of verbal explanations from the author.

has shown that *after the separation* the gradual condensation of the matter can make no difference whatever to the gravitational force exerted by the sphere on its surroundings; so that, if the universe outside the condensation was originally in equilibrium, its equilibrium will remain undisturbed. In these conditions the formation of condensations will start neither contraction nor expansion of the universe as a whole.

The actual conditions differ from the foregoing in that no empty crack is formed, the condensation merging gradually into its surroundings. The crack, by isolating the condensation from its surroundings, would have prevented any pressure of one on the other; in the absence of a crack there will be a pressure (probably exceedingly small) which will change as the condensation proceeds. It is this change of pressure, neglected in the preceding paragraph, that is the possible cause of expansion or contraction; for we have seen that the mere rearrangement of matter in a more condensed form has no effect. Lemaître describes the change which occurs as a "stagnation" of energy. It is not difficult to see that it is really the converse kind of change to that which occurs when energy of constitution of matter is liberated as radiation; energy is taken away from the transmissible form (pressure) and immobilised in the constitution of the condensation. Its effect is therefore opposite to that of conversion of material mass into radiation, and it tends to make the universe expand.

Sen's procedure is different. Having found the equations for a "pimply Einstein world" in static equilibrium, he calculates the total mass of such a world, and finds that it is always greater than that of a uniform Einstein world. It follows that, if the matter of the original uniform Einstein world is rearranged in condensations, there is not quite enough mass to form an equilibrium distribution. If we could artificially add a little mass to each condensation, we should obtain one of Sen's pimply spheres in equilibrium; the absence of this mass leaves the gravitational attraction in defect of the amount required to maintain equilibrium. Consequently cosmical repulsion has the upper hand and the universe expands.

Although both Lemaître and Sen agree that expansion (not contraction) ensues, there is a discrepancy between them; for Sen obtains it as the direct result of the rearrangement of matter, whereas Lemaître claims that the direct result is *nil*, and that the expansion is an indirect result dependent on the existence of a small pressure in the primordial nebula. Lemaître's investigation has the advantage that it avoids a very tricky calculation of the mass of the condensations, and seems to offer less likelihood of error.

It is only at the very beginning that we have to look for a cause of expansion or contraction; once started, the expansion or contraction continues and increases automatically. If there were causes of contraction and causes of expansion, victory went to the one which got its shove in first. Thus the formation of condensations must have had the start of the conversion of mass into radiation, since the latter would (as we have seen) have brought about a contracting universe. To my mind this rather suggests that the primordial material consisted of hydrogen (or equivalently free protons and electrons) since there would then be less opportunity for the conversion of mass into radiation than if more complex atoms were present. So long as they are not combined in complex nuclei, protons and electrons are immune from annihilation. The reason for this security is that the photon or quantum of radiation, which results from the annihilation of a proton and electron, has to be provided with momentum, which must be balanced by a recoil momentum. But in hydrogen there is nothing left to recoil. Annihilation of a proton and electron (if it ever occurs) can happen only when they form part of a complex system which will leave a residuum to carry the recoil.*

VI

We have been led almost inevitably to the consideration of the beginning of the universe, or at least to the beginning of the present order of physical law. This always happens when we treat of an irreversible one-way process; and the continual expansion of the world raises the same kind of question of an ultimate beginning as has been raised by the continual increase of entropy in the world.

Views as to the beginning of things lie almost beyond scientific argument. We cannot give scientific reasons why the world should have been created one way rather than another. But I suppose that we all have an aesthetic feeling in the matter. The solar system must have started somehow, and I do not

* I am indebted to Sir Alfred Ewing for calling my attention to this.

know why it should not have been started by projecting nine planets in orbits going in the same direction round the sun. But we have a feeling that that is not the way in which it would naturally be done; and we turn in preference to attempts—none too successful—to account for it by evolution from a nebula. Similarly the theory recently suggested by Einstein and de Sitter, that in the beginning all the matter created was projected with a radial motion so as to disperse even faster than the present rate of dispersal of the galaxies,* leaves me cold. One cannot deny the possibility, but it is difficult to see what mental satisfaction such a theory is supposed to afford.

Since I cannot avoid introducing this question of a beginning, it has seemed to me that the most satisfactory theory would be one which made the beginning *not too unaesthetically abrupt*. This condition can only be satisfied by an Einstein universe with all the major forces balanced. Accordingly the primordial state of things which I picture is an even distribution of protons and electrons, extremely diffuse and filling all (spherical) space, remaining nearly balanced for an exceedingly long time until its inherent instability prevails. We shall see later that the density of this distribution can be calculated; it was about one proton and electron per litre. There is no hurry for anything to begin to happen. But at last small irregular tendencies accumulate, and evolution gets under way. The first stage is the formation of condensations ulti-

* They do not state this in words, but it is the meaning of their mathematical formulae.

mately to become the galaxies; this, as we have seen, started off an expansion, which then automatically increased in speed until it is now manifested to us in the recession of the spiral nebulae.

As the matter drew closer together in the condensations, the various evolutionary processes followed— evolution of stars, evolution of the more complex elements, evolution of planets and life. Doubtless in this as in other theories there are serious difficulties of timing, so that one process should not go too fast compared with another. These difficulties of time-scale will be mentioned again later.

Perhaps it will be objected that, if one looks far enough back, this theory does not really dispense with an abrupt beginning; the whole universe must come into being at one instant in order that it may start in balance. I do not regard it in that way. To my mind *undifferentiated sameness* and *nothingness* cannot be distinguished philosophically. The realities of physics are unhomogeneities, happenings, change. Our initial assumption of a homogeneous static medium is no more than a laying out in order of the conceptions to be used in our analytical description of the distinguishable objects and events whose history we are going to relate. So far as these realities are concerned, the theory achieves its aim of providing an imperceptible and gradual beginning. When at last, by the thermodynamic degradation of energy, the universe with the same gradualness again reaches undifferentiated sameness, that is the end of the physical universe. I do not picture a worn out world careering forlornly through the rest of eternity. What is left is only a few conceptions which we forgot to put away after we had finished using them.

To illustrate the instability of an Einstein universe I will liken it to a pin standing on its point, which may fall either to the left or to the right into two horizontal positions *A* or *B*. Position *A* corresponds to a universe expanded to infinity, and position *B* to a universe contracted to a point or as nearly to a point as quantum conditions allow. As the only way of avoiding an abrupt beginning, I have supposed the pin to be vertical initially. Its balance then is not quite so precarious as it seems; it would be at the mercy of the slightest disturbance from outside—but there is *nothing* outside. So the fall must come from a slight "decay" in the material of the pin. According to Lemaître and Sen the decay is such as to make it fall towards *A*, and we now observe it midway in the fall.

If we do not mind a sudden, or even violent, beginning, many other experiments with the pin are possible. We may drop it from an inclined position, or in letting go give it a projection upwards or downwards. Starting from the horizontal position *B*, we may project it so that it rises and falls again; or, if projected with greater force, it may pass through the vertical position and fall on the other side into position *A*. Similarly if it is projected from *A*. The behaviour of a universe is precisely the same; to every adventure of the pin there corresponds a similar adventure of a universe and *vice versa*. These adventures have been treated at length by some writers, and the appropriate formulae calculated. Whilst such a mathematical study is proper in its own sphere, it is liable to give a misleading impression of the complexity of the problem before us. When the different projections are enumerated and presented as though they were all different "theories" of the universe, it looks as though we had come across a bewildering maze of possibilities. But all it amounts to is that the universe is just like any other system that has a position of unstable equilibrium.

At first sight there is a curious difference between the universe and the pin. If the universe has a given mass we cannot project it just how we please; in fact the circumstances of projection determine its mass. But this is explicable when we recollect that energy and mass are equivalent. The total energy of the pin varies according to the way it is projected, and strictly speaking its mass changes in the same way. The mass of the universe behaves analogously. To suppose that velocity of expansion in the (fictitious) radial direction involves kinetic energy, may seem to be taking our picture of spherical space too literally; but the energy is so far real that it contributes to the mass of the universe. In particular a universe projected from *B* to reach *A* necessarily has greater mass than one which falls back without reaching the vertical (Einstein) position.

Lemaître does not share my idea of an evolution of the universe from the Einstein state. His theory of the beginning is a *fireworks theory*—to use his own description of it. The world began with a violent projection from position *B*, i.e. from the state in which it is condensed to a point or atom; the projection was strong enough to carry it past the Einstein state, so

59

that it is now falling down towards *A* as observation requires. This makes the mass of the universe somewhat greater than in my theory (as explained in the last paragraph); but the change is scarcely important at the present stage of our progress. I cannot but think that my "placid theory" is more likely to satisfy the general sentiment of the reader; but if he inclines otherwise, I would say—"Have it your own way. And now let us get away from the Creation back to problems that we may possibly know something about".

The Einstein configuration was the one escape from an expanding or contracting universe; by proving it to be unstable, we show that it is no more than a temporary escape. Whether the original state was Einstein equilibrium or not, at the date when astronomers arrive on the scene they must be faced with an expanding or contracting universe. This result makes the theory of the expanding universe much more cogent. In 1917 theory was at the cross-roads (p. 1); that is no longer true, and by its own resources it has been guided into the road to a non-static universe. Realising that some degree of expansion (or contraction) is inevitable, we are much more inclined to admit the recession of the spiral nebulae as an indication of its magnitude.*

* I may mention that the proof of the instability of the Einstein configuration was the turning point in my own outlook. Previously the expanding universe (as it appeared in de Sitter's theory) had appealed to me as a highly interesting possibility, but I had no particular preference for it.

VII

Several counter theories of the observed recession of the nebulae have been proposed and I would like to make clear my general attitude to such theories.

I am a detective in search of a criminal—the cosmical constant. I know he exists, but I do not know his appearance; for instance I do not know if he is a little man or a tall man. Naturally the first move of my chief (de Sitter) was to order a search for footprints on the scene of the crime. The search has revealed footprints, or what look like footprints—the recession of the spiral nebulae. Of course, I am tremendously interested in this possible clue to the criminal. From the length of the stride I calculate the presumed height of the criminal (in approved detective fashion). Having gained this important information as to his appearance, I can now turn to my other clues—in relativity and wave-mechanics—and checking one against the other I think I have now about enough evidence to justify an arrest.

It happens that there are other persons interested in the footprints, who are not in the least interested in my criminal. For instance there is a geologist who suggests the theory that they belong to a prehistoric creature. (The counter theories proposed by Einstein and de Sitter and by Milne suppose that the large velocities of the nebulae have existed from the beginning.) Another man thinks they are not footprints at all, but depressions caused by something of unknown nature. To what extent is it incumbent on me to justify myself by criticising these contrary opinions?

I do not think they concern me at all closely. Naturally from the beginning I was awake to the possibility that the footprints might not belong to the criminal; the question then to be decided was not whether the clue was sufficient evidence to hang the criminal, but whether it indicated a direction of inquiry which it would be worth while devoting one's energies to following up. Of course, if either the geologist or the depressionist claimed to be able to demonstrate that his idea of the origin of the footprints was correct, I should pay grave attention; for such a demonstration would show that I was altogether on the wrong tack in my own inquiries. But that is not the position; no one claims more for the counter suggestions than that "for all we know, it might be so". That leaves the investigation as open as when we started: footprints have been discovered on the scene of the crime; all sorts of explanations are possible, and it may turn out that they are of little importance; but there is quite a good chance that they were made by the criminal; let us follow up the clue, and try to find out. I am fairly satisfied now that they do belong to the criminal, but that is because by pursuing the clue the further evidence detailed in Chapter IV has come to light.

I have already commented on the theory that the recession of the spiral nebulae is a misinterpretation of the red shift of their light. We may class together the remaining theories which accept the recession of the spiral nebulae as genuine; these accordingly admit the expansion of the universe (perhaps only as a temporary phenomenon) but do not connect it with cosmical repulsion.

The keynote of many of these suggestions seems to be, What is the most general deduction that can be made from our observational knowledge of the positions and motions of the galaxies? I think that those who seek this extreme generality are following a will-o'-the-wisp. The observational data give only the positions and velocities at the present instant; so that it is clear from the start that nothing can definitely be deduced as to the law of force governing the motion. Any *instantaneous* distribution of velocities is compatible with any law of force. If then anyone proposes to treat the problem of the system of galaxies with wider generality·than we here attempt—as he would perhaps say, without any preconceptions—we have to ask, What problem? The motions in themselves do not constitute a problem. We have to combine them with other ideas, which we think justified, in order to create a problem at all. It is the preconceptions—imported from other branches of science—that can fertilise an investigation otherwise doomed to barrenness.

Thus I find a difficulty in discussing the proposal of Einstein and de Sitter, and some of de Sitter's separate proposals, because I do not see what are "the rules of the game". These proposals are left as mathematical formulations, all doubtless compatible with what we observe; but there seems nothing to prevent such formulations being indefinitely multiplied. De Sitter has several times emphasised the possibility that the cosmical constant λ might be negative. This gives cosmical attraction instead of cosmical repulsion. Clearly the recession of the nebulae is not evidence in favour of cosmical attraction.

The most that can be said is that it is not necessarily fatal evidence against it.

It should not be forgotten that an observational test which is quite inadequate to demonstrate a theory may yet afford welcome confirmation of it. Suppose that by theoretical reasoning we have concluded that the earth is surrounded by a field of force attracting bodies towards it. To test this we are allowed one brief glimpse of what is happening near the earth's surface. Our glimpse may reveal a display of rockets soaring upwards. This is not incompatible with our theory, but it is clearly no confirmation of it. On the other hand we may see a shower of raindrops falling. Nothing can strictly be deduced from this one glimpse; but to observe objects falling to the ground is a tolerable confirmation of the theoretical prediction that there is a force tending to make objects move that way.

E. A. Milne* has pointed out that if initially the galaxies, endowed with their present speeds, were concentrated in a small volume, those with highest speed would by now have travelled farthest. If gravitational and other forces are negligible, we obtain in this way a distribution in which speed and distance from the centre are proportional. Whilst accounting for the dependence of speed on distance, this hypothesis creates a new difficulty as to the occurrence of the speeds. To provide a moderately even distribution of nebulae up to 150 million light-years' distance, high speeds must be very much more frequent than low speeds; this peculiar anti-Maxwellian distribution of

* *Nature*, July 2, 1932.

speeds becomes especially surprising when it is supposed to have occurred originally in a compact aggregation of galaxies.

I might discuss these suggestions more fully if they were likely to be the last. But it would seem that, unless we keep to a defined purpose, an unlimited field of speculation is open; and by the time these remarks are read, some other hypothesis may be in vogue. I define my own purpose as being to find what light (if any) the recession of the spiral nebulae can throw on the problem of the cosmical constant. Having regard to this purpose, it seems sufficient to note that this is not the only direction in which we might look for the explanation of the phenomenon of the nebulae, and then proceed with our task.*

* Further reference to the rival theories is made on p. 86.

Chapter III

FEATURES OF THE EXPANDING UNIVERSE

The world's a bubble, and the life of man
Less than a span.   Francis Bacon

I

A SPHERICAL world, closed but continually expanding, is a new playground for thought. Let us
play in it a little to familiarise ourselves with it. In
this chapter I shall mix together results which may
prove to be of scientific importance and results that
are probably no more than mathematical curiosities.
The plan is to set down anything that seems worthy
of note, even though we cannot see that it has any
ultimate importance in nature.

For a model of the universe let us represent spherical
space by a rubber balloon. Our three dimensions of
length, breadth and thickness ought all to lie in the
skin of the balloon; but there is only room for two, so
the model will have to sacrifice one of them. That
does not matter very seriously. Imagine the galaxies
to be embedded in the rubber. Now let the balloon
be steadily inflated. That's the expanding universe.

The galaxies are supposed to be scattered more or
less evenly over the surface; our observational knowledge, however, is limited to a portion which corresponds roughly to the size of France on a terrestrial
globe. The galaxies have individual motions, i.e.
motions with respect to the material of the balloon,

but these are comparatively small; in the main they recede from one another simply by the stretching of the rubber. The balloon, like the universe, is under two opposing forces; so we may take the internal pressure tending to inflate it to correspond to the cosmical repulsion, and the tension of the rubber trying to contract it to correspond to the mutual attraction of the galaxies, although here the analogy is not very close. Initially there was a balance; but a disturbance caused a slight expansion. This thinned out the rubber and made it less able to resist expansion. The more it expanded the less opposition was offered to expansion. The balloon is now probably several times its original size, and the tension of the rubber has decreased so much that it does little to retard the expansion.

A certain amount of quantitative data as to the dimensions, etc., of the universe can be obtained, and these I give forthwith. The figures are not final, but I think that (*a*), (*b*), (*c*), (*d*) are not likely to be in error by more than a factor 2 and the other two results by a factor 4:

(*a*) Speed of recession of distant objects (full value if the mutual attraction of the galaxies is negligible) = 528 kilometres per second per megaparsec distance.

(*b*) Initial radius of the universe before it began to expand = 328 megaparsecs = 1068 million light-years.

(*c*) Total mass of the universe = $2 \cdot 14 \cdot 10^{55}$ gm. = $1 \cdot 08 \cdot 10^{22} \times$ sun's mass.

67

(*d*) Number of protons in the universe = number of electrons = $1 \cdot 29 . 10^{79}$.

(*e*) Initial mean density of matter in the universe = $1 \cdot 05 . 10^{-27}$ gm. per cu. cm. = 1 hydrogen atom per 1580 cu. cm.

(*f*) The cosmical constant (λ) = $9 \cdot 8 . 10^{-55}$ cm.$^{-2}$

These results are interrelated; when one of them is known the others can all be deduced accurately. Thus they all depend on the value 528 which we here adopt as the speed of recession. From the observed speeds of recession of the spiral nebulae values ranging from 450 to 550 have been published. Strictly speaking the observed speed should be increased in order to obtain the "full value" referred to in (*a*), because we want to free the result from the drag of gravitational attraction; but making the best estimate we can of the masses of the nebulae, we judge that their mutual gravitational attraction is not likely to make an important difference. Many astronomers would adopt a value higher than 550, believing that Hubble's scale of distances of the nebulae is systematically too great. For this reason it would not be very surprising if the true value of the constant were as high as 1000 km. per sec. per mp.

The value 528 which is here used was adopted for the theoretical reasons discussed in Chapter IV. It depended on a preliminary development of the theory, and I can now see that it will be modified (probably increased) in the final theory. I might perhaps make a better shot at the value now; but it seems undesirable to chop and change whilst the theory is still incom-

plete. Thus at present both observational and theoretical values are subject to some uncertainty. However, since 528 is nearly the lowest value suggested, we
shall presumably not be exaggerating the effects of
the expansion if we adopt it.

The original radius of the universe is given under
(b), but we are unable to calculate the present radius.
It is rather tantalising not to know so important a
quantity; and unfortunately there is not much prospect of knowing it. I have a faint hope that some day
it will be revealed to us by the cosmic rays, if these
really are extra-terrestrial (see p. 80). Otherwise the
only method is to estimate the average density of
matter throughout the universe, and compare it with
the initial density given under (e); since the mass
cannot have changed importantly, the comparison
will give the expansion of volume and hence the expansion of radius. To find the present density we
should have to count the average number of galaxies
in a given volume, compute the average number of
stars per galaxy and the average mass of a star, and
allow also for the diffuse matter within the galaxies
and the still more diffuse matter between the galaxies.
I am afraid such an estimate can scarcely be trusted
to a factor of 100. The result, however, seems to come
out well below the value 10^{-27} found for the initial
density.

There is a curious difference between measuring the
radius of curvature of the expanding universe and
measuring the radius of the earth's surface. The
earth's radius gives no trouble provided that geodetic
measurements extend over a large enough area. It

might therefore be thought that our difficulty in measuring the present radius of the universe is due to the very small area of our survey, and will be removed when the survey is sufficiently extended. But the analogy fails because, owing to the delay of light-messages from distant parts of the universe, the information they can bring us is so much out of date that it would be useless as a guide to the *present* radius.

When occasion arises I shall assume for illustration that a 5-fold expansion of the original radius has occurred; but this number is merely a guess.

The mass of our own galaxy is roughly estimated at from 10^{10} to 10^{11} times the sun's mass. The average galaxy appears to be smaller. From the total mass of the universe given in (*c*) we conclude that there is enough material for at least a hundred thousand million galaxies.

A curious difficulty arises in stating the number of electrons or protons in the universe. Even if we count them one by one there is not a unique result, because there are two ways of counting, and one way gives twice as many as the other. It cannot be said that either way is wrong. According to one view, when we have counted the particles in one hemisphere of the spherical world we have finished the count, and the other hemisphere only gives us the same particles over again.* When we take this view we are said to use *elliptical* space (though the name "elliptical" does

* Take a long narrow strip of blotting-paper with a number of blots on it and form it into a ring with a twist in it. If you proceed continuously along the surface counting the blots you will after a time find yourself counting the same blots over again from the other side.

not seem very appropriate). It does not really matter which view we take, provided that we adhere to one view consistently. I adopt the other view throughout, and count both hemispheres of spherical space.

For those who care to examine the interrelation of these results, I add the leading mathematical formulae by which they are derived. The volume of a spherical space of radius R is $2\pi^2 R^3$.

This is larger than the ordinary Euclidean volume of a sphere. It is to be remembered that spherical space is not a Euclidean sphere but the skin of a four-dimensional hypersphere.

We call the initial (Einstein) radius R_e and the total mass M. These are related to the cosmical constant by

$$\lambda = 1/R_e^2 \qquad GM/c^2 = \tfrac{1}{2}\pi R_e,$$

G being the constant of gravitation ($6 \cdot 66 \cdot 10^{-8}$) and c the velocity of light. These results were obtained by Einstein in 1916.

The distance round the world is $2\pi R$, as though it formed a circuit of radius R—though physically the bending must be regarded as a fictitious representation and we get "round the world" literally by going straight on. It is only in an average sense that R is the radius of curvature; if we look at the universe microscopically the empty regions have less curvature and the regions occupied by matter have more curvature. A special importance is attached to the radius of curvature of the *empty* space, R_s. This is given by

$$R_s = R_e \sqrt{3} \qquad \lambda = 3/R_s^2.$$

The radius of curvature of the empty regions remains

constant, whilst the average radius expands. Thus the above formula for R_s holds at all times and not merely in the initial equilibrium condition. This to some extent compensates us for our lack of knowledge of the present radius of the universe; we do know (approximately) the present radius of curvature of any empty region.

The limiting speed of recession of the spiral nebulae or other distant objects is

$$c/R_s \text{ per unit distance.}$$

This is attained when the expansion has proceeded so far as to reduce the mutual attraction of the nebulae to insignificance. If q is the expansion, i.e. the ratio of the present radius to the original radius, the formula which takes account of the mutual gravitation is

$$\frac{c}{R_s}\sqrt{\left(1 - \frac{3}{q^2} + \frac{2}{q^3}\right)}.$$

Thus when $q = 2$, the speed of recession is 70·7 per cent. of the limiting value; when $q = 5$, it is 94·7 per cent. The formula is due to Lemaître.

The precise definition of R_s is "the radius of spherical curvature of any three-dimensional section of the four-dimensional continuum of space and imaginary time". For a non-expanding space it is the radius of curvature as ordinarily pictured; but the geometry of expanding space is too complex to afford a comprehensible picture of it. It might be thought that when the universe expands its radius can never become larger than the radius of curvature of empty space; but this refers to a different component of curvature from that measured by R_s.

II

One of the differences originally noticed between the Einstein and de Sitter universes was that light could go right round an Einstein universe but not round a de Sitter universe. Our expanding universe is on its way from the one condition to the other, and we ask how light will behave in it.

In the course of the expansion there is a definite moment after which circumambulation ceases to be possible. It seems certain that we are well past this moment, so that a ray of light emitted now will never get round to its starting-point again. On the other hand, light, which we now see, was emitted in the past. It may therefore have been emitted before the critical moment and have gone round the world before reaching us. Thus ideally it is still possible for us to see round the world; the events so seen must all have happened before the critical moment.

This may seem mysterious, but the mystery largely disappears when we remember that even at the beginning, when the circumference was at its smallest, light took 6700 million years to go round. Great changes are now happening in that period; the universe is doubling its radius every 1300 million years. Light is like a runner on an expanding track with the winning-post receding faster than he can run.

In the earliest days, when the universe was only just disturbed from equilibrium and the rate of expansion was slow, light and other radiation went round and round the universe until it was absorbed. This merry-go-round lasted until the universe had expanded to 1·003 times its initial radius. Then the bell rang for the last lap; light waves then running will make just one more circuit during the rest of eternity; those which started later will never get round.

Somewhat later, when the expansion reached 1·073,* the last half-lap began. After that moment it became impossible for light to travel half-way round; so that corresponding to any star or system there is a region of the universe which its light can now never reach. And if light cannot, no other causal influence can reach it, for no kind of signal can travel faster than light. I have sometimes pictured spherical space as a bubble. Our expanding universe is an expanding bubble. It seems fair to say that when the expansion reached 1·073 *the bubble burst.* For regions between which no causal influence can ever pass are as disconnected as the fragments of a bubble.

As I have already said, it is still quite possible for us to see things in or through the regions which are now broken off from us, because there is a lag of light-messages. In that case what we see refers to a time before the bursting of the bubble; the light got across before the breach occurred.

As light travels in the expanding universe it becomes reddened. Lemaître has shown that the reddening follows a simple rule, viz. the wave-length is increased in the ratio of the radius of the universe at the time of observation to the radius at the time of emission of the

---

* The critical values 1·003 and 1·073 were worked out by de Sitter. The critical moments are later in "elliptical" space (p. 70), because the runners then take a short cut leaving out one hemisphere.

light. For light which has been half-way round the world or more this reddening is considerable. We have seen that such light must have been emitted before the expansion reached 1·073, so that the radius at the time of emission was not much different from the Einstein radius. If the expansion is now 5, the wave-length will be increased 5-fold; this would shift most stellar spectra almost wholly into the infra-red. It has been suggested that a nebula seen in one direction in the sky might be the "back" of a nebula seen in the opposite direction. Apart from general unlikelihood, the extreme reddening of one or other image spoils this entertaining conjecture.

I should explain that the Doppler shift to the red due to the recession of the source of light is the *same* as the reddening here described. We have to explain it in different words, because we are now contemplating the passage of light over much greater distances. If the earth were to expand, the voyage between any two ports would be lengthened, and a transatlantic company might raise its fares on the ground that New York had receded from Liverpool; but for a round-the-world tour the statement that Liverpool had receded from Liverpool, however justifiable, would scarcely be an illuminating explanation. The reddening of light, like the raising of fares, is attributed sometimes to recession and sometimes more directly to the expansion, according to circumstances.

If you are in a spherical universe and look out in any direction, then if there is no obstruction you ought to see—the back of your head. Well, not exactly. The light has taken more than 6000 million years to go round, and your head was not there so long ago; but you ought to see what was in the position now occupied by your head. You will not need to use a powerful telescope to see it, because in fact you already have the most powerful telescope imaginable exactly adjusted on it, an instrument of some thousand million light-years' aperture. For this purpose spherical space is an optical instrument, neither lens nor mirror, but equally effective in bringing rays to a focus.

I am afraid, however, that you will not actually see it, even if you can get rid of all obstruction. A large telescope is not always a good telescope; its performance depends on the surfaces of the lenses or mirrors being perfectly true. The actual universe was not intended for an optical instrument, and it is not a true enough sphere to give any noteworthy results.

This brings us to the "theory of ghosts"—an idea developed more as a mathematical curiosity than as a serious physical speculation. In a perfectly spherical world rays of light emitted in all directions from a point will after travelling round the world converge to the same point; thus a real image is formed from which light will again diverge in all directions. Such an image might optically be mistaken for a substantial body. Owing to the time taken in circumambulating the world the image is not formed until at least 6000 million years later than its source. Other images would be formed after two circuits, three circuits, etc. We can thus imagine space to be populated not only with real stars and galaxies but with ghosts of stars which existed 6000 million, 12000 million, etc. years ago.

That would be a good practical joke for nature to play on the astronomer, but I do not think he would be taken in. The ghosts, being formed by light that has been round the world, would be ruddier than the genuine article, and so easily distinguished. Moreover, the trick could only be performed in the ideal spherical world of the mathematician; the actual universe is too irregular to focus the rays after their circuit. But the idea has another kind of interest because it raises an important question as to the meaning of relativity.

Considering a perfectly spherical transparent world, it is evident that the ghost is formed at the place occupied by the star when the light was emitted, not at the place now occupied by the star. Indeed the star may have since been scattered into many fragments. Thus if we observe simultaneously a star and its ghost, there will in general be a considerable distance between them due to the motion of the star during the circuit of the light. A star which coincides with its ghost has not moved. This gives us an observational test of "absolute rest" which appears to contradict the principle of relativity. We must admit that a perfectly spherical universe provides a frame of reference for distinguishing rest and motion, space and time, which can be looked upon as ultimate and absolute. It may be asked, Is not this precisely what the absolutists have been contending? Even if it is, I do not think they score very heavily by the admission that an absolute frame would exist if the universe were different from what it actually is. The perfectly spherical world is a mathematical invention, and any slight deviations from the sphere will cause the ghost to diffuse, so that absolute rest again becomes an indefinite conception. In a perfectly spherical universe nothing ever happens; for it is the irregularities that constitute the events. Even the expansion of its radius (by cosmical repulsion) means nothing, for there is no standard with which to compare it. Thus our admission that there can be an absolute time is coupled with the proviso that nothing ever happens in it.

Just as a frame of space and time defined with reference to the sun is appropriate for dealing with problems relating to the solar system, so a frame defined with reference to the matter of the universe as a whole is appropriate for dealing with the universe as a whole. In a uniform spherical world the frame appropriate to the universe as a whole is also appropriate to every part of it; thus the usual multiplicity of frames of space and time is suppressed. The principle of relativity is that one man's frame is as good as another's; it is not upset by imagining an ideal world in which every man has hit on the same frame: and to imagine circumstances in which there would be no opportunity for applying the principle is a very different thing from denying its validity.

In this book I speak of space and time as entirely distinct, and treat simultaneity as uniquely defined; it is to be understood that I am using a system of reference given by the universe as a whole, the universe having for this purpose been smoothed in the same way that in geodesy the earth is smoothed into the geoid. I claim no more for this frame than that it is convenient; in particular simultaneity (as defined by it) has no particular philosophical significance.

III

I have been speaking of the propagation of light, since it is the most familiar kind of radiation; but circumambulation becomes more of a practical possibility if we consider highly penetrating radiation, particularly the cosmic rays which are believed to come into our atmosphere from outside. The mean density of matter in the initial state of the universe and the length of the world-circumference are known (p. 67); hence it is easy to calculate that the average amount of obstruction to a cosmic ray in going round the world is equivalent to 7 cm. of water. It is well known that the rays can penetrate a much greater depth, so that it is possible for them to go many times round. It would seem that the cosmic rays generated almost from the beginning of time are still travelling through space, only a relatively small loss having occurred by absorption.

This is in keeping with the observed symmetry of their distribution, which otherwise seems inexplicable. Astronomical interest in cosmic rays was first aroused by the researches of Kolhörster; at that time it was stated that they were observed to come predominantly from directions in the plane of the Milky Way. Accordingly they were supposed to originate in the gaseous nebulae and diffuse matter occurring in our galaxy. The later and more accurate work of Millikan has proved, however, that there is no such galactic preference, and the distribution is approximately uniform in all directions. If then they have an extraterrestrial origin the source must be distributed symmetrically round the earth. But proceeding outwards from the earth, astronomy does not reveal anything with the necessary symmetry until we take in the whole universe. We might perhaps found on the cosmic rays an argument in support of closed spherical space; in an unclosed system it would be a strange coincidence if the earth were so centrally placed as to receive the rays equally from all directions.

The rays are generally supposed to be of sub-atomic origin, and attempts have been made to infer the exact origin from the wave-length (which is deduced by mathematical theory from the penetrating power). Just as the origin of a light ray is inferred from its wave-length, so the sub-atomic process which originated a cosmic ray might be inferred. But in these attempts it has been assumed that the observed wave-length is the original wave-length of the rays. One is reminded of the American in *Innocents Abroad* who refused to take any interest in a mummy and wanted to be shown a "nice fresh corpse". Those who are studying cosmic rays have taken it for granted that they are dealing with a nice fresh corpse, and are identifying the body on that supposition. I suggest that it is not a fresh corpse but a very ancient mummy. That will alter the identification considerably. If the cosmic rays date mainly from the relatively long period when the universe was close to its initial Einstein state, the wave-length has now increased perhaps 5-fold or more.

If I am right, cosmic radiation is a museum—a collection of relics of remote antiquity. These relics are stamped with an inscription indicating the dimensions of the world in its earliest ages. Whoever ultimately identifies the sub-atomic process originating the rays will be able to read the inscription and tell us just how much the universe has expanded since then.

The intensity of cosmic radiation falling on the earth is found to be not much less than the intensity of starlight (i.e. the radiation from all the stars except the sun). The earth, however, is in a privileged position with regard to starlight, being in the midst of one of the galaxies; it has no advantage with regard to cosmic rays which are presumed to have spread more or less evenly over all space. Thus in a more average region starlight is much less intense than cosmic radiation; and in a general survey of radiation in the universe only the cosmic radiation need be considered. It can be verified that the mass and energy of cosmic radiation throughout the universe is small compared with the mass of matter, and does not affect materially the formulae and calculations at the beginning of this chapter.

It is sometimes asked, What becomes of all the radiation which is continually being poured into space? A favourite speculation is that in some way it is ultimately turned back into matter. This has never seemed to me very likely, and it is discouraged by a conclusion recently pointed out by de Sitter. The total amount of radiation in the universe (measured by mass or by energy) is actually decreasing. Omitting for the moment changes by emission and absorption, the number of quanta remains constant; but the gradual increase of wave-length already discussed means that the energy of each quantum diminishes;

in fact the energy of the radiation diminishes in just the same proportion as the radius of the universe expands. This rate of diminution is now much too rapid to be compensated by fresh accessions from the radiation poured out by stars and by nebulous matter.

We may mention here a result which, though it refers to an entirely different topic, is closely related mathematically.* We do not expect the rule that the receding velocity of a spiral nebula is proportional to its distance to be obeyed by every individual. The rule refers to averages, and the nebulae will have individual velocities superposed on the systematic velocity of recession. It has sometimes been thought surprising that the individual deviations are not more conspicuous. Theory, however, shows that with the lapse of time the average individual velocity decreases in the same proportion as the universe expands; for example, if the expansion is now 5-fold, the nebulae have on the average only one-fifth of the individual velocities that they originally possessed.

IV

If the expanding universe is accepted, its most immediate reaction is on the time-scale of evolution. Three different time-scales have been favoured at one time or another, which we may distinguish as "short", "intermediate" and "long". No one now has a good word for the short Kelvin time-scale, which allowed

* The connection is that both results assert the decrease of a form of energy as the universe expands. They are comprised in a general theorem of Lemaître, that when the universe expands the pressure changes adiabatically. This is applied both to radiation pressure and to gas pressure, the "gas" in this case having galaxies for molecules.

not more than 50 million years for the whole life-time of the sun and solar system; and practically our choice lies between the intermediate scale giving the sun an age of the order 10^{10} years, and the long scale which gives its age fairly definitely as $5 \cdot 10^{12}$ years. Naturally it is the policy of the evolutionist to grab as much time as possible in order to give his processes a longer opportunity to accomplish something. So when there is no strong evidence one way or another, the longer time-scale gets the preference. This, rather than any striking success of the theory, accounts for the popularity of the long time-scale in recent years.

It is sometimes forgotten that of the two hypotheses, "long" and "intermediate", the long time-scale is the older. It came about through Einstein's theory which gave the total amount of energy in any given mass of matter. Knowing the mass of the sun, we knew just how much energy it contained and could calculate how long this would maintain the present rate of radiation if it could all be released. If *all* the energy of constitution of matter is released, the matter must disappear; so that it is necessary that the electrons and protons should annihilate one another. At the time of my first researches on the internal constitution of the stars (1916–1917) the annihilation of electrons and protons was the only source of energy known that was adequate for stellar requirements, and it was the only one that I then discussed; but in 1920 a possible alternative was recognised in the energy released by the transmutation of hydrogen into higher elements. This releases rather less than 1 per cent. of the total amount of energy contained in the matter, so that it gives by no means so abundant a supply as the total annihilation of matter. Still it suffices for the intermediate time-scale, and we could perhaps make it do at a pinch.

There is no direct evidence that the annihilation of protons and electrons can occur—unless we count the evidence of the cosmic rays, which according to some authorities are supposed to contain a wave-length which indicates this source. If the long time-scale could be established by astronomical researches it would be good indirect evidence, for it seems clear that there is no way of providing for it without annihilation of matter. Direct evidence for the building up of complex elements out of hydrogen scarcely seems to be required; since the elements exist, there is presumably a way of forming them. But it may be remarked that the recent discovery of the neutron makes it much easier to envisage the steps of this process.

I do not think that anything very decisive has been found for or against either theory (annihilation of matter or transmutation of hydrogen) or for or against either time-scale (long or intermediate). Like other time-grabbers I have generally adhered to the long time-scale provisionally, since it affords more scope for investigation. But two years ago I was much shaken by a study of the dynamics of our Milky Way system; its form and construction seem to be such that it is impossible that it should have endured for the period of the long time-scale.*

* Halley Lecture, *The Rotation of the Galaxy* (Oxford Univ. Press 1930).

In a universe doubling its radius every 1300 million years, it is evident that the long time-scale of billions of years is altogether incongruous. It is true that our theory does not set any definite limit to past time. There may have been a very long period of approximate equilibrium before any serious expansion began; but this scarcely counts from the point of view of stellar evolution. Astronomical history may be said to begin when the first condensations were fully formed and the galaxies separated from one another; but by this time the expansion must have been well under way. It is difficult to allow much more than 10^{10} years between then and now.

Thus astronomers, who have been luxuriating in an enormously long time-scale, are threatened with a drastic cut. Even in these days of economy, a cut of about 99 per cent. is not to be accepted lightly by the department concerned. I confess that I do not quite see how we are going to manage on the reduced allowance; and I am not disposed to blame those whose reaction is to try to seek for some loophole by which the cut can be avoided.

If we find it hard to accept the speed at which the universe is changing, acceptance is not made easier by the consideration of what it is changing towards. The fragments of the burst bubble will continually become more numerous until each galaxy is a separate fragment. I suppose that the distance of one galaxy from the next will ultimately become so great, and the mutual recession so rapid, that neither light nor any other causal influence can pass from one to another. All connection between the galaxies will be broken;

each will be a self-contained universe uninfluenced by anything outside it. Such a disintegration is rather a nightmare to conceive; though it does not threaten any particular disaster to human destiny (see p. 89).

But the difficulty as to the time-scale does not arise from the more recondite parts of the present theory. It remains whether we adopt spherical space or not. It becomes *more serious* if we do not accept the relativity explanation of the recession. Since the outward speed of the nebulae is known by observation, it is a matter of simple arithmetic to compute the date when they were close together in a congested crowd. It was not more than 1900 million years ago. This assumes that in the past the speeds were as great as they are now. The relativity explanation grants some reprieve, for it postulates that the outward speeds have been gradually produced by cosmical repulsion and were therefore smaller in the past. The rival explanations, which do not admit a repulsive force, refuse this extension, and consequently accentuate the difficulty of the time-scale.

So I cannot see anything to be gained by discarding the relativity explanation in favour of one of the other theories. On the one point in which the present results appear to offer some difficulty, the other theories greatly magnify the difficulty. I have much more sympathy with those critics who deny the nebular recession altogether, believing the observed radial velocities to be spurious (p. 16). They are relieved of all difficulty as to the time-scale.

Speaking on the de Sitter-Lemaître theory in September 1931, I said—"But the theory of the expanding universe is in some respects so preposterous that we naturally hesitate to commit ourselves to it. It contains elements apparently so incredible that I feel almost an indignation that anyone should believe in it—except myself".

The reasons for the exception are dealt with in Chapter IV.

V

Expansion is a relative term. When we assert that the universe expands, what do we assume to be constant? This question is often asked, but there is no subtlety about the answer. The expansion is relative to the standards of length ordinarily accepted—the metre bar for example. The wave-length of cadmium light is sometimes adopted as the best available standard of length; expressed in terms of the cadmium wave-length, the radius of the universe is continually increasing.

By the same standard the radius of curvature of an *empty* region of space remains constant (p. 72). We can easily see that whilst a surface as a whole is swelling the curvature of certain portions of it can remain unchanged.

The fact that the expansion is relative to ordinary standards is clear enough when we adopt the standpoint of Chapter I; but the more recondite treatment, in which the phenomenon is presented as a uniform expansion of spherical space, seems to have confused not only the casual reader but some of the experts. Their idea is that, since every part of space is being inflated uniformly, the inflation will affect distances between electrons in the atom and between planets in the solar system just as much as distances between galaxies. This would mean that atoms, human beings, the earth, the solar system, expand at the same rate as the universe; there would be no change in the radius of the universe expressed in metres, since the metre bar is expanding at the same rate. Such an expansion shared by everything alike would be undetectable, and would in fact have no definable meaning.

The fallacy arises from forgetting that the expanding spherical universe is a very much simplified model. We cannot appeal to it to decide how atoms, measuring rods and planets behave, because atoms, measuring rods and planets have been smoothed away into a perfectly continuous and uniform distribution of mass. The inflation is only uniform if the density is uniform. If we consider a roughened or pimply sphere, it is found mathematically that the roughened parts do not expand at the same rate as the smooth intervals between them.

Lemaître designed his expanding spherical space for the treatment of phenomena affecting the universe as a whole. His approximation is grotesquely inadequate for treating smaller scale phenomena such as the behaviour of measuring rods or the internal structure of a galaxy. Within a galaxy the average world-curvature is some thousands of times greater than Lemaître's average for the universe as a whole, and his formulae are inapplicable.

The result is that only the intergalactic distances expand. The galaxies themselves are unaffected; and all lesser systems—star clusters, stars, human observers and their apparatus, atoms—are entirely free from expansion. Although the cosmical repulsion or expansive tendency is present in all these smaller systems, it is checked by much larger forces and no expansion occurs. To see how this happens, suppose that the sun and planets are given rather large electrical charges of the same sign; that would introduce an expansive tendency into the solar system, but it would not turn it into an expanding system. After an initial readjustment the planets would describe periodic orbits as before in the modified field of force, and the solar system would not grow any larger. This holds until the charge on the planets is made so strong that the repulsion outweighs gravitation; the planets then abandon the periodic type of orbit and recede continually. Thus the demarcation between permanent and dispersing systems is quite abrupt. It corresponds to the distinction between periodic and aperiodic phenomena.

It appears then that the "bursting of the bubble" will end when each galaxy is a separate fragment. It will not go on to disrupt the galaxies. These no doubt contain their own seeds of decay, and cosmical repulsion may ultimately help to scatter their fragments; but that concerns a much more distant future. If you think that the shattering of the bubble universe is a tragic outlook, it may be some consolation to reflect that when the worst has happened our galaxy of about a hundred thousand million stars will be left intact. It is not so bad a prospect.

VI

All change is relative. The universe is expanding relatively to our common material standards; our material standards are shrinking relatively to the size of the universe. The theory of the "expanding universe" might also be called the theory of the "shrinking atom".

It is our instinctive outlook that *we* are always the same; it is our environment that changes. As with Anatole France's dog Ricquet—"Les hommes, les animaux, les pierres grandissent en s'approchant et deviennent énormes quand ils sont sur moi. Moi non. Je demeure toujours aussi grand partout où je suis".

Is not the expanding universe another example of distortion due to our egocentric outlook? Surely the universe should be the standard and we should measure our own vicissitudes by it. We see a relative change, and cry out that the universe is dissolving; as well might the growing child, who sees the familiar home becoming smaller, be dismayed at the vanishing property of houses and furniture.

The argument sounds plausible, but I do not deem it true. Even if our standards are held responsible for the expanding of the universe, they cannot be held responsible for its bursting. Moreover our constant standards are not necessarily puny. I have mentioned (p. 72) one cosmical dimension which remains constant, namely the radius of curvature R_s of empty regions of the universe. Since it stands in a constant ratio to the metre, it can be used equivalently. This is in fact the ideal cosmical standard, and judged by it the universe changes whilst we remain true to size.

Although I do not think the suggestion goes very deep or that it has any philosophical moral, I will follow it for my last escapade in our new playground. Let us then take the whole universe as our standard of constancy, and adopt the view of a cosmic being whose body is composed of intergalactic spaces and swells as they swell. Or rather we must now say it keeps the same size, for he will not admit that it is he who has changed. Watching us for a few thousand million years, he sees us shrinking; atoms, animals, planets, even the galaxies, all shrink alike; only the intergalactic spaces remain the same. The earth spirals round the sun in an ever-decreasing orbit. It would be absurd to treat its changing revolution as a constant unit of time. The cosmic being will naturally relate his units of length and time so that the velocity of light remains constant. Our years will then decrease in geometrical progression in the cosmic scale of time. On that scale man's life is becoming briefer; his threescore years and ten are an ever-decreasing allowance. Owing to the property of geometrical progressions an infinite number of our years will add up to a finite cosmic time; so that what we should call the end of eternity is an ordinary finite date in the cosmic calendar. But on that date the universe has expanded to infinity in our reckoning, and we have shrunk to nothing in the reckoning of the cosmic being.

We walk the stage of life, performers of a drama for the benefit of the cosmic spectator. As the scenes proceed he notices that the actors are growing smaller and the action quicker. When the last act opens the curtain rises on midget actors rushing through their parts at frantic speed. Smaller and smaller. Faster and faster. One last microscopic blurr of intense agitation. And then nothing.

Chapter IV

THE UNIVERSE AND THE ATOM

See Mystery to Mathematics fly!
Pope, *Dunciad*

I

I HAVE explained in the previous chapters that theory led us to expect a systematic motion of recession of remote objects, and that by astronomical observation the most remote objects known have been found to be receding rapidly. The weak point in this triumph is that theory gave no indication how large a velocity of recession was to be expected. It is as though an explorer were given instructions to look out for a creature with a trunk; he has brought home an elephant—perhaps a *white elephant*. The conditions would equally well have been satisfied by a fly, with much less annoyance to his next-door neighbour the time-grabbing evolutionist. So there is great argument about it.

I think the only way to remove the cloud of doubt is to supplement the original prediction, and show that physical theory demands not merely a recession but a particular speed of recession. The theory of relativity alone will not give any more information; but we have other resources. I refer to the second great modern development of physics—the quantum theory, or (in its most recent form) wave-mechanics. By combining the two theories we can make the desired theoretical calculation of the speed of recession.

93

This is a new adventure, and I do not wish to insist on the accuracy or finality of the first attempt. I cannot see how there can be anything seriously wrong with it; but then one never does see these faults until some new circumstance arises or some ingenious person comes along to show us how blind we have been. But there are two kinds of scientific misadventure; we may start off on a false trail altogether, or we may make temporary blunders in following the true path. I am content if in this chapter I can justify my belief that at any rate we are not committing the first error.

According to the argument here developed we can calculate by pure theory what ought to be the speed of recession of the spiral nebulae. (This is subject to the reservation that the restraining effect of their mutual gravitational attraction is relatively unimportant, a condition which appears to be satisfied in the present state of the universe.) Since certain small factors in the formulae are at present left in suspense, there is a temporary indefiniteness; but we can say provisionally that the result is between 500 and 1000 km. per sec. per megaparsec. No astronomical observations of any kind are used in this calculation, all the data being found in the laboratory. Therefore when we turn our telescopes and spectroscopes on the distant nebulae and find them to be receding at a speed within these limits the confirmation is striking.

The original prediction of de Sitter and Lemaître gave no indication whether the phenomenon would first become perceptible at nebular distances or at distances 10^6 or 10^{60} times greater. We had not the faintest idea how large an effect would appear. By the new investigation, however, the amount is so closely defined that there can be little doubt as to the correspondence of the theoretical and observed effects. Our astronomical explorer cannot be accused of having brought home an elephant in mistake for a fly; and (if I may further complicate the zoological metaphor) even if it is a white elephant it is not a mare's nest.

Any theoretical step requires testing in as many directions as possible. If the theoretical ideas here employed had had only one application, viz. to calculate the recession of the nebulae, there might be a certain amount of room for "fudging". As a matter of fact the danger of unconscious fudging is greatly exaggerated; there is an artistry in these fundamental equations of physics which one cannot trifle with. But it naturally strengthens our confidence if the same step also leads to the solution of another problem. This happens in the present case, the associated problem being the relation of the proton to the electron and in particular the ratio of their masses. Here a very delicate observational test of the theory is possible.

Thus we are not dealing with an isolated problem but with a theory which determines at the same time two of the leading constants of physics, viz. the *cosmical constant* and through it the recession of the nebulae, and the *mass-ratio* of the proton and electron.

I cannot give here the mathematical part of the argument. I want rather to show that all the necessary physical ideas present themselves naturally, and are waiting for the mathematician to express them in symbols and work out the answer. By a preliminary attempt at the latter task we gain fair assurance that no serious difficulty is likely to arise.

II

We have been contemplating the system of the galaxies—phenomena on the grandest scale yet imagined. I want now to turn to the other end of the scale and look into the interior of an atom.

The connecting link is the cosmical constant. Hitherto we have encountered it as the source of a scattering force, swelling the universe and driving the nebulae far and wide. In the atom we shall find it in a different capacity, regulating the scale of construction of the system of satellite electrons. I believe that this wedding of great and small is the key to the understanding of the behaviour of electrons and protons.

You will see from the formulae on p. 71 that the cosmical constant is equal in value to $1/R_e^2$ or to $3/R_s^2$, so that it is really a measure of world-curvature; and in place of it we can consider the initial radius of the universe R_e, or better the steady radius of curvature of empty space R_s. In the present chapter the unqualified phrase "radius of curvature" or the symbol R will be understood to refer to R_s. Being the radius *in vacuo* it has the same kind of pre-eminence in physical equations that the velocity of light *in vacuo* has. I will first explain why the radius of curvature is expected to play an essential part in the theory of the atom.

Length is relative. That is one of the principles of Einstein's theory that has now become a commonplace of physics. But it was a far from elementary kind of relativity that Einstein considered; according to him length is relative to a frame of reference moving with the observer, so that as reckoned by an observer moving with one star or planet it is not precisely equal to the length reckoned by an observer moving with another star. But besides this there is a much more obvious way in which length is relative. Reckoning of length always implies comparison with a standard of length, so that length is relative to a comparison standard. It is only the ratio of extensions that enters into experience. Suppose that every length in the universe were doubled; nothing in our experience would be altered. We cannot even attach a meaning to the supposed change. It is an empty form of words —as though an international conference should decree that the pound should henceforth be reckoned as two pounds, the dollar two dollars, the mark two marks, and so on.

In *Gulliver's Travels* the Lilliputians were about six inches high, their tallest trees about seven feet, their cattle, houses, cities in corresponding proportion. In Brobdingnag the folk appeared as tall as an ordinary spire-steeple; the cat seemed about three times larger than an ox; the corn grew forty feet high. Intrinsically Lilliput and Brobdingnag were just the same; that indeed was the principle on which Swift worked out his story. It needed an intruding Gulliver—an extraneous standard of length—to create a difference.

It is commonly stated in physics that all hydrogen atoms in their normal state have the same size, or the same spread of electric charge. But what do we mean by their having the same size? Or to put the question the other way round—What would it mean if we said that two normal hydrogen atoms were of different sizes, similarly constructed but on different scales? That would be Lilliput and Brobdingnag over again; to give meaning to the difference we need a Gulliver.

The Gulliver of physics is generally supposed to be a certain bar of metal called the International Metre. But he is not much of a traveller; I do not think he has ever been away from Paris. We have, as it were, our Gulliver but have left out his travels; and the travels are, as Prof. Weyl was the first to show, an essential part of the story.

It is evident that the metre bar in Paris is not the real Gulliver. It is one of those practical devices which serve a useful purpose, but dim the clear light of theoretical understanding. The real Gulliver must be ubiquitous. So I adopt the principle that when we come across the metre (or constants based on the metre) in the present fundamental equations of physics, our aim must be to eject it and to substitute the natural ubiquitous standard. The equations put into terms of the real standard will then reveal how they have arisen.

It is not difficult to find the ubiquitous standard. As a matter of fact Einstein told us what it was when he gave us the law of gravitation $G_{\mu\nu} = \lambda g_{\mu\nu}$. Some years ago I showed that this law could be stated in the form, "What we call a metre at any place and in any direction is a constant fraction $(\sqrt{\tfrac{1}{3}\lambda})$ of the radius of curvature of space-time for that place and direction". In other words the metre is just a practically convenient sub-multiple of the radius of curvature at the place considered; so that measurement in terms of the metre is equivalent to measurement in terms of the radius of curvature.

The radius of world-curvature is the real Gulliver. It is ubiquitous. Everywhere the radius of curvature exists as a comparison standard indicating, if they exist, such differences as Gulliver found between Lilliput and Brobdingnag. If we like we can use its sub-multiple the metre, remembering, however, that the metre is ubiquitous only in its capacity as a sub-multiple of the radius. We should, if possible, try to forget that in certain localities we have crystallised this metre into metallic bars for practical convenience.

We can now give a direct meaning to the statement that two normal hydrogen atoms in any part of the universe have the same size. We mean that the extent of each of them is the same fraction of the radius of curvature of space-time at the place where it lies. The atom here is a certain fraction of the radius here, and the atom on Sirius is the same fraction of the radius at Sirius. Whether the length of the radius here is absolutely the same as that of the radius at Sirius does not arise; and indeed I believe that such a comparison would be without meaning. We say that it is always the same number of metres; but we mean no more by that than when we say that the metre is always the same number of centimetres.

Thus it appears that in all our measures we are really comparing lengths and distances with the radius of world curvature at the spot. Provided that the law of gravitation is accepted, this is not a hypothesis; *it is the translation of the law from symbols into words.* It is not merely a suggestion for an ideal way of measuring lengths; it reveals the basis of the system which we have actually adopted, and to which the mechanical and optical laws assumed in practical measurements and triangulations are referred.

It is not difficult to see how it happens that our practical standard (the metre bar) is a crystallisation of the ideal standard (the radius of curvature, or a sub-unit thereof). Since the radius of curvature is the unit referred to in our fundamental physical equations, anything whose extension is determined by constant physical equations will have a constant length in terms of that unit. Thus the physical theory that provides that the normal hydrogen atom shall have the same size in terms of the radius of curvature wherever it may be, will also provide that a solid bar in a specified state shall have the same size in terms of the radius of curvature wherever it may be. The fact that the atom has a constant size in terms of the practical metre is a case of "things which are in a constant ratio to the same thing are in a constant ratio to one another".*

The simplification obtained by using the actual radius of curvature as unit of length (instead of using a sub-unit) is that all lengths will then become angles in our world-picture. The measure of any length will be the "tilt of space" in passing from one extremity to the other. It is true that these angles are not in actual space but in fictitious dimensions added for the

* For a fuller explanation see *The Nature of the Physical World*, Chapter VII.

purpose of obtaining a picture; but the justification of the picture is that it illustrates the analytical relations, and these angles will behave analogously to spatial angles in the mathematical equations.

To sum up this first stage of our inquiry: If in the most fundamental equations of physics we adopt the radius of curvature R_s as unit instead of the present arbitrary units, we shall have at least made the first step towards reducing them to a simpler form. We know that many equations are simplified when velocities are expressed in terms of the velocity of light *in vacuo*; we expect a corresponding simplification when lengths are expressed in terms of the radius of world-curvature *in vacuo*. When the equation is in this way freed from irrelevant complications it should be easier to detect its true significance. We cannot make this change of unit so long as the ratio of R_s to our ordinary unit is unknown; but observation of the spiral nebulae has provided us with what we provisionally assume to be an approximate value of R_s, so that it is now possible to go ahead with our plan.

III*

In elementary geometry we generally think of space as consisting of infinitely many points. We approach nearer to the physical meaning of space if we think of it as a network of distances. But this does not go far enough, for we have seen that it is only the ratios of distances which enter into physical experience. In

---

* This section is mainly an additional commentary on the principles explained under II. If found too difficult, it can be omitted. The main argument is resumed in IV.

order that a space may correspond exactly to physical actuality it must be capable of being built up out of ratios of distances.

The pure geometer is not bound by such considerations, and he freely invents spaces consisting only of points without distances, or spaces built up out of absolute distances. In adapting his work for application to space in the physical universe, we have to select that part of it which conforms to the above requirement. For that reason we must reject his first offer—flat space. Flat space cannot be constructed without absolute lengths, or at least without a conception of *a priori* comparability of lengths at a distance which can scarcely be distinguished from the conception of absolute length.*

Flat space, being featureless, does not contain within itself the requirement for reckoning length and size, viz. a ubiquitous comparison standard. But what

* In pre-relativity theory, and in the original form of Einstein's theory, "comparison of lengths at a distance" was assumed to be axiomatic; that is to say, there was a real difference of height between the Lilliputian and the Brobdingnagian irrespective of any physical connection between the islands. The fact that they were in the same universe—phenomena accessible to the same consciousness—had nothing to do with the comparison. Such a conception of unlimited comparability is scarcely distinguishable from the conception of absolute length. In a geometry based on this axiom, space only does half its proper work; the purpose of a field-representation of the relationships of objects is frustrated, if we admit that the most conspicuous spatial relationship, ratio of size, exists *a priori* and is not analysable by field-theory in the way that other relationships are. Weyl's theory rejected the axiom of comparability at a distance, and it was at first thought that such comparability could not exist in his scheme. But both in Weyl's theory and in the author's extension of it (affine field-theory) it is possible to compare lengths at a distance, not as an extra-geometrical *a priori* conception, but by the aid of the field which supplies the ubiquitous standard necessary.

is the use of a space which does not fulfil the functions of space, namely to constitute a scheme of reference for all those physical relations—length, distance, size—which are counted as spatial? Since it does not constitute a frame of reference for length, the name "space" is a misnomer. Whatever definition the pure geometer may adopt, the physicist must *define* space as something characterised at every point by an intrinsic magnitude which can be used as a standard for reckoning the size of objects placed there.

No question can arise as to whether the comparison unit for reckoning of lengths and distances is a magnitude intrinsic in space, or in some other physical quality of the universe, or is an absolute standard outside the universe. For whatever embodies this comparison unit is *ipso facto* the space of physics. Physical space therefore cannot be featureless. As a matter of geometrical terminology features of space are described as curvatures (including hypercurvatures); as already explained, no metaphysical implication of actual bending in new dimensions is intended. We have therefore no option but to look for the natural standard of length among the radii of curvature or hypercurvature of space-time.

To the pure geometer the radius of curvature is an incidental characteristic—like the grin of the Cheshire cat. To the physicist it is an indispensable characteristic. It would be going too far to say that to the physicist the cat is merely incidental to the grin. Physics is concerned with interrelatedness such as the interrelatedness of cats and grins. In this case the "cat without a grin" and the "grin without a cat" are equally set aside as purely mathematical phantasies.

When once it is admitted that there exists everywhere a radius of curvature ready to serve as comparison standard, and that spatial distances are directly or indirectly expressed in terms of this standard, the law of gravitation $(G_{\mu\nu} = \lambda g_{\mu\nu})$ follows without further assumption; and accordingly the existence of the cosmical constant λ with the corresponding force of cosmical repulsion is established. Being in this way based on a fundamental necessity of physical space,* the position of the cosmical constant seems to me impregnable; and if ever the theory of relativity falls into disrepute the cosmical constant will be the last stronghold to collapse. *To drop the cosmical constant would knock the bottom out of space.*

It would be a truism to say that space is not an ultimate conception; for in the relativity view of physics every conception is an intermediary between other conceptions. As in the closed universe described in Chapter II, where the galaxies form a system having no centre and no outside, so the conceptions of physics link into a system with no boundary; our goal is not to reach an ultimate conception but to complete the

---

* The requirement is that the comparison standard shall be a magnitude intrinsic in the space—for whatever the standard is intrinsic in, that *ipso facto* is space. Space can have other characteristic magnitudes besides the radius of curvature—for example, magnitudes measuring various kinds of hypercurvature. Although the suggestion seems far-fetched, it is, I suppose, conceivable that one of these might be substituted. That would give a different law of gravitation; but there is still a cosmical constant, depending on the ratio of the metre to the natural comparison standard. In fact the cosmical term $\lambda g_{\mu\nu}$ remains unchanged; it is $G_{\mu\nu}$ which is modified.

full circle of relationship. We have concluded that the ubiquitous comparison standard must be a characteristic of space, because it is the function of space to afford such a standard; but we can inquire further how space and the standard contained in it themselves originate.

The space in which the atom is pictured as having position and size is an intermediary conception used to relate the atom to the "rest of the universe". It is therefore no contradiction if we say sometimes that the extension of the atom is controlled by the curvature of space, and sometimes that it is controlled by forces of interaction proceeding from the rest of the universe. It must be remembered that we are only aware of an atom or any other object in so far as it interacts with the rest of the universe, and thereby gives rise to phenomena which ultimately reach our senses. The position and dimensions which we attribute to an atom are symbols associated with interaction effects; for there is no meaning in saying that an atom is at A rather than at B unless it makes some difference to something that it is at A not B. In considering this interaction it is not necessary to deal separately with every particle and every element of energy in the rest of the universe; if it were, progress in physics would be impracticable. For the most part it is sufficient to take averages. The multitudinous particles of the universe admit of an almost uncountable variety of change of configuration; in considering their interaction with the atom we need preserve only a few broad types of average change. The "rest of the universe" is thus idealised into something possessing only a few types of variation or degrees of freedom. This is illustrated in electrical theory where the interactions of myriads of electrical particles are replaced by the interaction of an *electric field* which is specified uniquely by six numbers. In the same way another part of the interaction of the rest of the universe on the atom is idealised into interaction of a *metrical field*, or—to give it its usual name—*space*. The few broad types of variation which are not smoothed out by averaging are retained in the curvatures of space.

We must distinguish in conception between space which for certain purposes replaces the rest of the universe and space which is occupied by the rest of the universe, although the two spaces ultimately become identical. The distinction is easier if we use the term "metrical field" instead of "space"; for (by analogy with electrical fields) we recognise that a field has a dual relation to matter, viz. it is produced by matter and it acts on matter.

The remainder of our task is to try to discover the details of this idealisation of the "rest of the universe" into a metrical field containing a radius of curvature.

IV

One of the most fundamental equations of physics is the wave-equation for a hydrogen atom, that is to say for a proton and electron. The equation determines the size of the atom or the spread of its electric charge. Clearly the ubiquitous standard of length R must come into this equation.

Now R does not appear in the equation as ordinarily written. That is because the equation has been reached through experiment, and is expressed in terms of quantities such as the charge of an electron, Planck's constant, the velocity of light, etc. The radius R though present is in disguise. We must try to penetrate the disguise.

At first sight a formidable obstacle appears. The radius of the hydrogen atom is of order 10^{-8} cm., and the natural unit R is of order 10^{27} cm.; thus the radius of the hydrogen atom in terms of the natural unit is of order 10^{-35}. Our idea was that by introducing the natural unit we should obtain a simplified equation; but can it be a very simple equation if its solution is 10^{-35}? Clearly it must contain an enormous numerical coefficient in one or more of its terms. If the equation is really in its most elementary form, every coefficient ought to have some simple meaning—some obviously appropriate reason for being what it is. We should not be surprised to see the 4π type of coefficient, which has a simple geometrical meaning; or a coefficient equal to the number of dimensions or degrees of freedom concerned in the problem, which arises from summing together a number of symmetrical terms. But what simple meaning can be attached to an enormously large number like 10^{35}?

I can think of only one large number which is in any way relevant to the problem, viz. the number of particles (electrons or protons) in the universe. Indeed there seems to be no other way of putting a large number into the structure of the physical world. I refer, of course, to pure numbers, not to the kind of number that we arbitrarily introduce by our centimetre-gram-second system of reckoning. We shall find presently that there are direct reasons for assuming that the number of particles in the universe N will occur in the coefficients of the wave-equation; but even without these reasons the enormous magnitude of the coefficients would be a sufficient indication of the occurrence of N.

Another aspect of the same large ratio appears when we compare the electric force between a proton and electron with the gravitational force between them. According to classical theory the ratio is $2 \cdot 3 \cdot 10^{39}$. I have long thought that this must be related to the number of electrons and protons in the universe* and I expect that the same view has been entertained by others. Since N is about 10^{79} (p. 68) the above ratio is of the order $\sqrt{N}$

The direct reason for the appearance of N is that N is actually an effective number of degrees of freedom of the universe. On classical theory the number would be greater than N, because each of the N particles would have several degrees of freedom; but there is a well-known exclusion principle which limits the freedom of a particle by forbidding it to go into an orbit already occupied by another particle. Wave-mechanics therefore approaches the problem from the other end and defines N as the number of independent wave-systems existing in the universe, and therefore equal to the number of separate constituents of the energy of the universe. It is quite possible that the number N approached in this way will be found, not to be arbitrary, but to have some definite theoretical foundation; but that is pure conjecture, and for the

* *Mathematical Theory of Relativity*, p. 167.

present we regard it as the one arbitrary element in the design of the actual universe.

Our atom is situated in and interacting with a universe containing N degrees of freedom. We idealise and simplify the problem by picturing it as situated in and interacting with a space (or metrical field) of radius of curvature R possessing a comparatively small number of degrees of freedom, say n. In this simplified form "the rest of the universe" comes into the equation of the hydrogen atom through the quantity R. I think we must expect that the numbers N and n will also occur in the equation, as a memorandum of the substitution of a space with n degrees of freedom for a universe with N degrees of freedom. For four-dimensional space-time the number n is found to be 10. We shall lose sight of it for the time being, but it will turn up later.

Having decided that N and R will enter into the coefficients of the equation of the hydrogen atom, we next inquire in what kind of association they will occur. The factor N, of course, arises from adding together equal contributions from each of the particles or wave-systems; the question is, What is the nature of the contributions to be added and how do they contain R? I do not profess to have achieved the necessary physical insight to settle the question. It will no doubt be much more satisfactory when we have a picture in which we can, as it were, see these entities adding themselves together, just as we can see a hundred centimetres adding themselves to form a metre. But when this kind of insight fails, we are not without a guide. As conduct may be guided by ethics or by "good form", so this kind of investigation can be guided by physical insight or by analytical form. Both wave-mechanics and relativity theory are *very strict on good form*. Only certain kinds of entities are allowed to be added together. To add anything else would be a solecism. "It isn't done."

In relativity theory the only things that are additive are action-invariants.* The action-invariant containing R is the Gaussian curvature, which is proportional to $1/R^2$. In quantum theory the entities which may be added are the squares of momenta, or as they are written symbolically $\partial^2/\partial x^2$. To construct a quantity of the same dimensions out of R we must take $1/R^2$.† I take it therefore that the entities to be added are, or are proportional to $1/R^2$; so that the required combination is N/R^2.

This gives us what may be called an "adjusted natural standard of length", viz. $R/\sqrt{N}$. By using $R/\sqrt{N}$ instead of R as our unit we absorb the factor N, so that it will not trouble us any more. From the data on p. 67 the length of the adjusted standard is about 3.10^{-13} cm., so that it is not unsuitable for dealing with phenomena of electrons.

Now we can go back to our problem, which was to discover how the natural standard of length is disguised in the familiar wave-equation. But this time

---

* Other tensors may only be added if they are at the same point of space—a condition which is obviously not fulfilled here.

† The guidance of quantum theory is less obvious than that of relativity theory because the former commonly adopts a mixed system of units (dynamical and geometrical). Relativity theory being purely geometrical avoids the complication.

we look for the adjusted standard $R/\sqrt{N}$ instead of the original standard R.

I think I have identified the adjusted standard in the wave-equation, disguised as the expression e^2/mc^2. Here e is the charge of an electron or proton, m the mass of an electron and c the velocity of light. This expression is well known to be of the dimensions of a length; in fact $\frac{2}{3}e^2/mc^2$ used to be called the "radius of an electron" in the days when the electron was conceived more substantially than it is now. The identification accordingly gives the equation

$$\frac{R}{\sqrt{N}} = \frac{e^2}{mc^2}.$$

I cannot enter here into the justification of this identification, which would lead deeply into the principles of quantum theory. But I may mention that the identification is a very simple one. The expression e^2/mc^2, or rather its reciprocal, stands rather disconsolately by itself in the wave-equation, forming a separate term. Investigators, who are busy transforming, explaining, theorising on the other terms, leave it alone; it has just been accepted as ballast. It calls out for identification.

It may be asked, Is not a straight identification too simple? Granting the identification in principle, will there not be a numerical factor—say $\frac{1}{2}$, or 2π, or perhaps something more complicated—the type of factor which usually appears when we reach the same entity by different routes? Perhaps there is; but at present the simple identification looks to me to be correct.* I should add, however, that I am uncertain whether in this formula N is to be taken as the number of electrons or the number of electrons *and* protons, so that a factor $\sqrt{2}$ is left in suspense. For definiteness I here take N to be the number of electrons only; the number of protons must be approximately, and probably exactly, the same.

By the relativity theory of the expanding universe, we have

$$\frac{N}{R} = \frac{\pi}{2\sqrt{3}} \frac{c^2}{Gm_p},$$

where m_p is the mass of a proton. This follows readily from the formulae given on p. 71; the only points to notice are that the notation R_s is here changed to R, and that the total mass M of the universe is approximately Nm_p.

Thus by relativity theory we find N/R, and by wave-mechanics we find $\sqrt{N}/R$. Combining the two results we find N and R separately. The resulting value of N is about 10^{79}. From R the limiting speed of recession of the galaxies c/R is found immediately. All the constants involved have been measured in the laboratory. The agreement of the result obtained in this way with the observed recession of the nebulae has already been described (p. 94).

The following summary of the theory is due to Prof. Dingle:

* Except that there are certain corrections amounting altogether to less than 1 per cent. which are explained under V. The mathematical arguments on which the identification is based are given in *Proceedings of the Royal Society*, vol. CXXXIII A, p. 605, and *Monthly Notices of the R.A.S.* vol. XCII, p. 3.

He thought he saw electrons swift
    Their charge and mass combine.
He looked again and saw it was
    The cosmic sounding line.
The population then, said he,
    Must be 10^{79}.

V

Having, as we think, detected the adjusted natural standard in the terms of the wave-equation, we have next to inquire how our result affects the theory of protons and electrons. For in identifying the standard with e^2/mc^2 we have taken a step which links the universe to the atom; and we ought to verify the observational consequences not only in 'the astronomical universe but also within the atom.

In wave-mechanics the momentum of a particle is usually stated to be

$$\frac{ih}{2\pi} \frac{\partial}{\partial x}.$$

The factor $h/2\pi$ is an unnecessary complication due to our haphazard choice of units of length and mass. We shall adopt instead a natural unit of mass, which is related to the unit of length in such a way that the momentum is simply $i\partial/\partial x$. The meaning of i (literally the square root of -1) in an equation is that the two sides of the equation represent waves which, though equal in amplitude, are a quarter-period different in phase. When the mass of an electron is expressed in terms of this natural unit we shall denote it by m_e. Making the change of unit, the identification on p. 111 becomes

$$\frac{hc}{2\pi e^2} m_e = \frac{\sqrt{\mathcal{N}}}{R} \qquad \ldots\ldots(A).$$

We have taken the opportunity to turn both sides upside down, since that is the way they actually appear in the wave-equation.

The coefficient $hc/2\pi e^2$, which is sometimes called the *fine-structure constant*, is a pure number; and it is well known that its value is close to 137. For my own part I think that its value is exactly 137, that being the number of degrees of freedom associated with the wave-function for a pair of charges. There has been much discussion whether the true value is 137·0 or 137·3; both values claim to be derived from observation. The latter, called the "spectroscopic value", is preferred by many physicists. It is, however, misleading to call these determinations *observational values*, for the observations are only a substratum; the spectroscopic value in particular is based on a rather complex theory and is certainly not to be treated as a "hard fact" of observation.

Although I believe $hc/2\pi e^2$ to be 137, I shall take the actual coefficient of m_e to be 136. This means that I slightly amend the original identification by inserting a factor 136/137. The reason for this change is that one of the 137 degrees of freedom, viz. that corresponding to radial displacement, occurs in some problems but not in others, and this appears to be a problem in which it will not occur. The fine-structure constant is introduced in the problem of interaction of two electric charges, and there the 137 degrees of freedom are all in play; a change of distance between the two charges is recognisable because a comparison standard for distances is furnished by the radius of curvature R. But now we are considering a formula for the mass of an electron, which arises from its inter-action with the "rest of the universe". The N particles of the universe have been virtually simplified down to one particle by introducing the adjusting factor $\sqrt{N}$, so that the problem is not dissimilar to that of two interacting particles; but there is no longer an ex-traneous comparison standard of length. Tracing the analogy between the two problems, we find that the analogue of change of distance between the two electrons would be change of the radius of curvature of space. But by its very nature R cannot vary, since it is the standard unit of distance. There is therefore no analogue to the 137th degree of freedom; and we conclude that our first identification, which did not enter into such minutiae, ought to be amended so as to show the correct number of degrees of freedom.

One might hesitate to introduce so odd-looking a factor as 136/137 were it not that we know of another case in which the radial degree of freedom is inhibited, and there the factor has been verified by observation. This occurs when a proton enters into the almost rigid helium nucleus. Its mass or energy is found to be reduced in a ratio which is very nearly 136/137; the reduction is called the *packing-fraction*. The disap-pearance of a degree of freedom is essentially the same in the helium nucleus and in the metrical field; the former cannot expand radially because it is rigid, the latter cannot expand because its radius is the standard of length.

Thus at present our result stands

$$136m_e = \frac{\sqrt{N}}{R} \qquad \ldots\ldots(B).$$

But to this there is a serious objection. The result shows an unfair discrimination in favour of the electron, the proton not being mentioned. The proton is presumably as fundamental as the electron. But what can we put in place of $\sqrt{N}/R$ which would give an equally fundamental equation for the mass m_p of a proton?

With an electron and proton calling out for equal treatment the only way to satisfy their claims impartially is to make the fundamental equation a quadratic, so that there is one root for each. We do not want to alter the part we have already got, after taking so much trouble to justify it bit by bit; so we assume that

$$136m - \sqrt{N}/R = 0 \qquad \text{......(C)}$$

gives correctly the last two terms of the equation, but there is a term in m^2 to come on at the beginning.

It is well known that we can learn something about the roots of a quadratic equation, even if only the last two terms are given. The ratio of the last two coefficients is the sum of the roots divided by the product of the roots. Since the equation is to have roots m_e and m_p, we must have

$$\frac{m_e + m_p}{m_e m_p} = \frac{136R}{\sqrt{N}}$$

or

$$\frac{136 m_e m_p}{m_p + m_e} = \frac{\sqrt{N}}{R} \qquad \text{......(D)}.$$

This is another change in the identification equation; but this time it is a very small change numerically. Comparing (D) with (B) we see that a factor $m_p \div (m_p + m_e)$ has been inserted. We know that m_p

is about 1847 times m_e, so that the factor is 1847/1848 or ·99946. Numerically the change is insignificant; but the proton no longer has any cause of complaint, for proton and electron receive perfectly impartial treatment in (D).

The next step is to complete the quadratic equation of which the last two terms are given in (C). Since we have finished with the problem of the identification of the adjusted standard (our final equation giving it in terms of known experimental quantities being (D)) we may as well now adopt it as our unit of length. As already explained this choice of unit ought to reduce the equations to their simplest possible form. This means that $R/\sqrt{N}$ can now be taken as unity. The two terms given in (C) are therefore $136m - 1 = 0$, and the completed quadratic is

$$? m^2 - 136m + 1 = 0.$$

What number must we put in place of the query? You may remember that there was a number $n = 10$, which we promised to bring into the wave-equation* sometime. Here is our chance. We take the equation to be

$$10m^2 - 136m + 1 = 0 \qquad \ldots\ldots(E).$$

For reference we write down the same equation, re-introducing the centimetre as the unit of length. It is

$$10m^2 - 136m \frac{\sqrt{N}}{R} + \frac{N}{R^2} = 0 \qquad \ldots\ldots(F).$$

You see that the number $n = 10$ occurs in the first term as a counterweight to N in the last term, which is evidently their proper relation.

* The wave-equation is formed by replacing the mass m by a differential operator.

Although we are not yet able to give a clear-cut theory of this equation, the argument is further advanced than might appear from this superficial sketch. I think it will be agreed that, since the coefficient 136 represents number of degrees of freedom, it is extremely probable that the remaining coefficient in (E) will also represent number of degrees of freedom. Presumably each degree of freedom possesses a concealed energy or cyclic momentum similar to that provided by the ignoration of co-ordinates in ordinary dynamics. Approaching the problem from this dynamical point of view one arrives almost immediately at the term $10m^2$, but there is more difficulty in seeing the reason for the term $136m$.

Thus, broadly speaking, we have two lines of approach. They have not yet met and coalesced, as they ultimately must do. But since one line of approach gives the linear term and the other the squared term, apparently quite definitely, the progress seems to be already sufficient to give the correct equation.

We find the ratio of the two roots m_p, m_e, by solving equation (E) or (F). The result is $1847 \cdot 60$. The observed value* of the mass-ratio of the proton

* There is also a spectroscopic value about 10 units lower, obtained by adopting $137 \cdot 3$ instead of 137 for the fine-structure constant; but this is irrelevant. We are already committed to the value 137 at an earlier stage, so we *do not want* our computation to agree with a result which would only be true if the value were not 137. To put it another way—Naturally our theory cannot agree with both deflection values and spectroscopic values, since these differ; but it ought to agree consistently with one set, and not sometimes with one and sometimes with the other.

and electron is stated to be 1847·0 with a probable error of half a unit. Thus the agreement is complete.

The solution of equation (F) gives

$$
\left.\begin{array}{l}
135{\cdot}9264\mathrm{m}_e = \sqrt{N}/R \\
0{\cdot}073569\mathrm{m}_p = \sqrt{N}/R
\end{array}\right\} \quad \ldots\ldots (G).
$$

This is another form of our final identification of the adjusted standard $R/\sqrt{N}$. The preliminary identification was $137\mathrm{m}_e = \sqrt{N}/R$; so that there has not been any change worth considering in the numerical magnitude of $\sqrt{N}/R$, derived for the purpose of predicting the speed of recession of the nebulae. The formulae will work either way. Normally we apply them to calculate the astronomical results, using m_e or m_p determined by physicists; but we can also use them to give an astronomical method of measuring the mass of an electron or proton.

To measure the mass of an electron, a suitable procedure is to make astronomical observations of the distances and velocities of spiral nebulae! The result, corrected if necessary for the mutual attraction of the galaxies is, let us say, 600 km. per sec. per megaparsec. This is c/R (p. 72); and since the velocity of light c is 300,000 km. per sec., we have $R = 500$ megaparsecs, which is equal to $1{\cdot}54 \ldotp 10^{27}$ cm. The remaining steps, which involve a little algebraic handling of the equations need not be described in detail; as soon as R is known, they become soluble and we can find N, and hence $\sqrt{N}/R$. The masses of the electron and proton are then given by equations (G). They are there expressed in terms of the natural unit of mass; we can convert them into grams if we know Planck's constant h. I am afraid that the accuracy attainable by this method would not satisfy the modern physicist, but we shall not be out by more than a factor 2 or thereabouts.

Perhaps you will object that this is not really *measuring* the mass of an electron; even supposing it to be right, it is a highly circuitous inference. But do you suppose that a physicist puts an electron in the scales and weighs it? If you will read an account of how the spectroscopic value of the mass has been determined, you will not think my method unduly complicated. But, of course, I do not seriously put it forward in rivalry; I only want to make vivid the wide interrelatedness of things.

We can show that the two roots of the quadratic represent electric charges of opposite sign. To test this an electric field must be introduced into the problem. Following Dirac's theory, this is done by adding to the equation a constant term (i.e. a term not involving the differential operator $m = id/ds$) depending on the electrical potential. Since this changes only the third term of the quadratic, the sum of the roots $m_e + m_p$ is unchanged. In other words, the mass or energy added by the field to m_e is equal and opposite to the mass or energy added by the field to m_p. But that is the definition of equal and opposite charges—in the same electric field they have equal and opposite potential energy.

Our conclusion that the fundamental wave-equation is really a quadratic has recently received unexpected support by the discovery of the neutron. It had been supposed that the wave-equation for two charges (a proton and electron*) was a linear equation first given by Dirac. The complete set of solutions was found to represent (approximately, if not exactly) a hydrogen atom in its various possible states. But from recent experiments it has been discovered that there exists another state or group of states in which the proton and electron are much closer together and form a very minute kind of atom. This is called a neutron. Clearly the present wave-equation for a proton and electron cannot be correct, since its solutions do not give the neutron states. Just as for one charge we require a quadratic wave-equation whose two sets of solutions correspond to electrons and protons, so for two charges we require a quadratic wave-equation whose two sets of solutions correspond to hydrogen atoms and neutrons. I expect that this continues in more complicated systems, the two solutions then corresponding to extra-nuclear and nuclear binding of the charges.

The support to the present theory is twofold. Firstly it indicates that the theory of two charges will come into line with our theory of one charge as regards the general form of the equation. Secondly, since the "spectroscopic values" of the physical constants are based on an incomplete theory of two charges, we should not attach overmuch importance to a slight discrepancy between them and the values found in our theory.

It is perhaps still more important that we can see other problems ahead. Thus it will be necessary to

* Note that our quadratic equation (E) is for a proton *or* electron—not for two charges.

investigate the theory of the additional term in the equation for the hydrogen atom and neutron, and to show that the extra solutions agree with the observed properties of the neutron. This will give further opportunities of testing, and if necessary revising, the present conclusions.

VI

To those whose interest in modern science is directed chiefly to the philosophical implications, the theory of the expanding universe does not, I think, bring any particularly new revelation. Except for one lapse, I have avoided questions savouring of philosophy. I have rather taken it for granted that the reader's attention, like my own, is fixed on the strictly scientific progress of the inquiry, and that he will suspend all questions as to how the physical scheme, which is here being developed, can be made to fit in with the general outlook of life and consciousness. It would be unfortunate to prejudice the inquiry by dragging in such questions prematurely.

We may perhaps emerge with the uneasy feeling that we no longer have vast domains of space and vast periods of time to dispose of. But the complaint has often been made that astronomical measures are too vast to conceive; and if we have never been able to conceive them, it scarcely affects our general outlook to have a few o's lopped off here and there. In fact I consider the man who is dissatisfied with a universe containing ten thousand million million million stars rather grasping. That is, if he wants them merely for comfortable philosophic contemplation; in connection with scientific investigations the cuts may, of course, be serious, as we have seen in the case of the time-scale.

The new theory contains no obvious suggestion that the world will come to an end sooner than we had been expecting. The cosmical dispersal ignores the smaller scale aggregations like our galaxy. We anticipate that there will ultimately be a complete running-down of the universe by the slow degradation of energy into unavailable form; but that far distant day is not brought noticeably nearer by the existence of cosmical repulsion.

It would seem that the expansion of the universe is another one-way process parallel with the thermodynamical running-down. One cannot help thinking that the two processes are intimately connected; but, if so, the connection has not yet been found.

The position with regard to the thermodynamical running-down of the universe has not materially altered since I discussed it four years ago.* The impression has got abroad that the conclusions have been shaken by recent work on cosmic rays. That would be impossible, so far as I am concerned; for the theory of cosmic rays that is being urged in this connection happens to be the one that I was advocating at the time of writing, viz. that the cosmic rays give evidence of the building up of higher elements out of hydrogen in distant regions occupied by diffuse matter.† I am not at all sure that the more recent evidence should be interpreted as favourable to it; but if it is favourable, as Dr Millikan maintains, I have

---

* *The Nature of the Physical World*, Chapter IV (1928).
† *Internal Constitution of the Stars*, p. 317 (1926).

the less reason to change my views. The coming together of electric particles to form a complex atom, and the consequent dispersal of some of the energy in cosmic rays, is clearly a step in the same direction as other energy-dissipating processes—for example, the coming together of nebulous matter to form a star, and the consequent dispersal of energy as radiant heat. It is one more contributor to the general running-down towards an ultimate state of thermodynamic equilibrium. Millikan has sometimes called the atom-building process a "winding-up" of the universe; but "up" and "down" are relative terms, and a transformation of axes may be needed in comparing his descriptions with mine.

It may be desirable to remind the philosophical reader of the reason why the scientist indulges in these extrapolations of our present imperfect knowledge to regions remote from our experience—why he writes about the beginning and end of the world. It seems to be gratuitously courting disaster to expose our theories to conditions in which any slight weakness is likely to become magnified without limit. But that is just the principle of testing. "The real justification for making such forecasts is not that they are likely to be realised; but that they throw light upon the state of contemporary science, and may indicate where it requires supplementing."*

The test of extrapolation to the most distant future does not, I think, disclose any definite weakness in the present system of science—in particular, in the second law of thermodynamics on which physical science so

* Prof. H. T. H. Piaggio.

largely relies. It is true that the extrapolation foretells that the material universe will some day arrive at a state of dead sameness and so virtually come to an end; to my mind that is a rather happy avoidance of a nightmare of eternal repetition. It is the opposite extrapolation towards the past which gives real cause to suspect a weakness in the present conceptions of science. The beginning seems to present insuperable difficulties unless we agree to look on it as frankly supernatural. We may have to let it go at that. But I have referred elsewhere to the danger of limiting scientific investigation to a bounded domain (p. 28). Instead of honestly facing the intricacies of our problem, we may be led to think that its difficulties have been solved when they have only been swept over the boundary. Sweeping them back and back, the pile increases until it forms an unclimbable barrier. Perhaps it is this barrier that we call "the beginning".

VII

Now I have told you "everything right as it fell out". How much of the story are we to believe?

Science has its showrooms and its workshops. The public to-day, I think rightly, is not content to wander round the showrooms where the tested products are exhibited; the demand is to see what is going on in the workshops. You are welcome to enter; but do not judge what you see by the standards of the showroom.

We have been going round a workshop in the basement of the building of science. The light is dim, and we stumble sometimes. About us is confusion and mess which there has not been time to sweep away.

The workers and their machines are enveloped in murkiness. But I think that something is being shaped here—perhaps something rather big. I do not quite know what it will be when it is completed and polished for the showroom. But we can look at the present designs and the novel tools that are being used in its manufacture; we can contemplate too the little successes which make us hopeful.

A slight reddening of the light of distant galaxies, an adventure of the mathematical imagination in spherical space, reflections on the underlying principles implied in all measurement, nature's curious choice of certain numbers such as 137 in her scheme —these and many other scraps have come together and formed a vision. As when the voyager sights a distant shore, we strain our eyes to catch the vision. Later we may more fully resolve its meaning. It changes in the mist; sometimes we seem to focus the substance of it, sometimes it is rather a vista leading on and on till we wonder whether aught can be final.

Once more I have recourse to Bottom the weaver—

I have had a most rare vision. I have had a dream,—past the wit of man to say what dream it was: man is but an ass, if he go about to expound this dream....Methought I was, and methought I had,—but man is but a patched fool, if he will offer to say what methought I had....

It shall be called Bottom's dream, because it hath no bottom.

Index

INDEX